THE SEA OF CONSCIOUSNESS

by Gerard Aartsen

featuring

THE INVISIBLE OCEAN

plus two articles

by George Adamski

"The universe, as we now know, is not a domain of matter moving in passive space and indifferently flowing in time; it is a sea of coherent vibrations."

–Professor Ervin Laszlo, philosopher and systems scientist, in *The Intelligence of the Cosmos* (2017)

George Adamski, circa 1930

Contents

Publisher's Foreword

It is a mark of the appeal of George Adamski's work that an increasing number of his books are being reprinted by various publishers today, even several that are still kept in print by the George Adamski Foundation itself.

This continued interest in Adamski's work is also reflected in the number of publications about the man, his experiences and his teaching that have appeared in the last ten years or so, although this still falls short of the steady interest that exists in Japan, where his collected writings are available in no less than 12 volumes.

The fact that Adamski's very first publication, released in 1932 and reproduced in the volume before you now, is not included in his collected writings and was until now unknown to even his most dedicated students, underlines how unexpected the discovery of *The Invisible Ocean* was.

Read in conjunction with two other, previously unpublished articles, also included here in Part I, *The Invisible Ocean* accentuates the fact that his interest and his teaching were consistent throughout his life and, if anything, only matured in their expression.

In Part II, we will look at some aspects of his earliest thought in the context of the latest insights from systems science – the scientific approach that looks at life, consciousness and the universe as interrelated parts of an integral whole.

When read in the light of systems science and quantum physics, which holds that nothing is separate, that everything interacts as an integral system which evolves as a whole, we see that George Adamski's philosophy and teaching were far ahead of their time and are now finding confirmation at the very forefront of 21st-century scientific thinking. This, of course, requires a complete reassessment

vii

of Adamski's entire body of work on the part of detractors, sceptics and critics – everyone who has thus far been convinced he was a conman, or who was less than convinced he could be speaking from a higher level of understanding than most of us have access to.

This volume, therefore, is of historical importance because it shows that George Adamski has always taught about the evolution of consciousness, the beckoning examples of which are the Masters of Wisdom with whom he trained in Tibet as a youth, and the Space Brothers whose philosophy of Life he described at length in his book *Inside the Space Ships*.

The short essays in this, his very first publication, also show that he always had a deep interest in Life as a space-based principle, and that he knew well to look beyond the carbon-bound manifestation of consciousness that is still the focus of all but the most advanced thinkers today.

We see it as a unique privilege to be able to make this small treatise on what George Adamski called Universal Law available again and give his many students around the world a glimpse of how his teaching has evolved, allowing the world a fuller understanding of the true significance of his mission, as a tribute to the visionary teacher that he was.

17 April 2019

Publisher's Note: Due to the poor quality of the copy of *The Invisible Ocean* at our disposal it was decided to re-compose the text for this reproduction *verbatim*, except for the rearranging of two lines of text that were transposed on the original page 6 and the correction of a few obvious spelling mistakes and inconsistencies.

Also, footnotes provide references to passages from the New Testament where applicable.

PART I

THE
INVISIBLE OCEAN

Author
Geo. Adamski

PREFACE

The Author's purpose of this booklet is to give out to readers and students the message that has been seeking expression for some time. It is a Message of Divine power, not draped with veils but stated in clear, simple word pictures. Now as never before, the subject of mystic power is attracting the attention of thinking people. It is time to make all things plain so that no one will be lost in the search for TRUTH. The illustration of the Ocean herein given will give the reader the reason for the many different states of consciousness in the one body called the universe or GOD, and its terrific power.

Professor George Adamski
████████ of Tibet.

PREFACE

The Author's purpose of this booklet is to give out to readers and students the message that has been seeking expression for some time. It is a Message of Divine power; not draped with veils but stated in clear, simple word pictures. Now as never before, the subject of mystic power is attracting the attention of thinking people. It is time to make all things plain so that no one will be lost in the search for TRUTH. The illustration of the Ocean herein given will give the reader the reason for the many different states of consciousness in the one body called the universe or GOD, and its terrific power.

Professor George Adamski of Tibet

1

FLOATING POWER IN THE INVISIBLE OCEAN

Do you realize that we are living in an invisible ocean? Do you realize that this invisible ocean has a tremendous power? It is more powerful than our visible oceans, which we have named Pacific, Atlantic, etc. Scientists today say that if they could harness the ocean, it would give power to the whole world, yet few of us know that we are now living in an invisible ocean of power like that. As above, so below.

We are actually walking in a sea of power, a power so tremendous that no man can measure it, or know its limits. If so powerful, you would ask, how is it that it does not destroy us? Because man is as strong as that power, and the inner man has as much power as the universe, and the temple in which he dwells has the same strength.

In time of storm should we try to buck this power we may be wrecked like a ship on the visible ocean. When we sway with the power, we are safe and the storm subsides. Ships drift with the waves in time of storm. They never try to buck them. The same principle holds good in the invisible ocean. We are swaying with the power when we say, "Thy will be done." Doing that no storm can destroy us.

What is this invisible ocean and what proof do have we that it exists? In order for the visible ocean to exist, it must

receive its essence from the invisible ocean. We believe that we have seven oceans, but were we to go down far enough and investigate, we would find out that all the [2]* oceans connect, and are really one. But even this one great ocean does not compare with the invisible ocean that is everywhere with its boundaries nowhere.

We know the invisible through the essence of the visible. Water is manifested by the descension of oxygen and hydrogen into a lower vibration. Hydrogen and Oxygen is the essence of water. Hydrogen and Oxygen is boundless and limitless.

When men drill for oil, they find two kinds of gas. They call one wet gas, and the other dry gas. The wet is the one manifesting on the earth plane as wet, while the other is still in a gaseous form and can be reduced to a wet condition. In the same way, we have dry and wet water. The dry would be the gas Oxygen and Hydrogen that is invisible in a high ratio vibration before it descends into the visible when it becomes wet water. So we have the wet and the dry, or the visible and the invisible ocean.

Let us consider the power of the visible ocean. We know no limits to the power of the visible ocean. Its depths are so vast that no man has measured it. Let us, for instance, place the depth of the ocean at 3,000 feet. In this great depth, there are various strata with different rates of pressure. The deeper we go down, the greater the pressure. When on the surface of the water, we find that our bodies can float. Down ten feet, the pressure increases, and as we continue to descend, the

* Numbers in square brackets indicate end of text on the original page with that number.

pressure gradually becomes greater and greater. We find, as we go down, that there are different vibrations in the ocean, for pressures are [3] vibrations. On the surface the vibrations are finer, as we descend the vibrations become coarser.

In this visible ocean which we are discussing, there are countless thousands of fish and plants. Some inhabit the floor of the ocean, and yet are not crushed by that pressure or vibration. Living a few feet above them is another creature that cannot perhaps live at the very bottom of the ocean. It must stay in its own strata or vibration. We can now see that between the floor of the ocean and the surface there are many stratas of different vibration and that different creatures inhabit each strata.

We can say, too, that in each different vibration there is a particular feeling or condition that each creature dwelling there partakes of. It does not feel so much the condition of the vibration above or below it. Yet because of an inner life urge resident in every living thing it is gradually drawn upward in its evolution. Inasmuch as there are layers upon layers in the ocean, there must be millions of different states of consciousness in which the ocean creatures are living, for what we call consciousness is a state of feeling. They feel this difference when they go out of their particular strata into another. The fish that lives at the bottom of the ocean cannot understand what is on the top. It is trying to obtain a higher consciousness but being unable to go higher, it does not as yet know the consciousness of the surface. Whatever partakes of that vibration lives in that certain strata. The fish

that finally advances to the highest strata finds himself on the surface and sees the light of the world* for the first time.

We humans are living in an invisible ocean at the [4] present time, most of us at the bottom of it. The only way we can graduate to the surface is to use our determination or rather the feeling urge we have within us, that there IS something above us. This feeling will finally get us to the surface for we shall strive to see what is there.

As we grow in consciousness, we get nearer and nearer the light and finally we shall see the full light, just as the fish can begin to see the light when it is within fifteen feet or so from the surface. It is the light just above us that draws us higher and higher, and makes us more and more determined to get to the surface. Some people become so determined to get there very quickly, that they tire themselves and thus cause delay. If we remain natural and easy going, we would get there just the same.

The natural way is the best way.

The fish that is able to go to the bottom of the ocean and then swim to the surface is the MASTER OF THE OCEAN. He can do what all other fish can do. When he swims to the surface of the ocean, he sees the light of the sun and the sun appears like a great God. He "goes in and out and finds pasture."† He begins to unfold through greater knowledge and observation. He goes to the bottom and tells the weaker ones all about the wonderful things he knows and has seen. Most of them will tell him he is crazy. Some will condemn him because they cannot conceive of anything higher than

* John 8:12, "Then spake Jesus again unto them, saying, I am the light of the world: he that followeth me shall not walk in darkness, but shall have the light of life."

† John 10:9, "I am the door: by me if any man enter in, he shall be saved, and shall go in and out, and find pasture."

their strata. But the master fish will find pasture there because some fish will believe and will want to grow into a master fish.

As the master fish passes through the different strata, [5] he will find fish that will appreciate his knowledge and teaching. As it is with the fish, so it is with us, only that we work on different planes of mental vibrations instead of strata of water. The master among us can go down to the depths of the earth and up to the lights of heaven. Nothing can hold him for he is above any condition. He can descend to the level of people in the coarser vibration in order to teach them about the light. He goes through layer by layer of consciousness, mixing freely with all the people and here and there he finds some who are ready to learn about the light. The same principle works in the visible and in the invisible ocean. We call the invisible ocean spirit, force, or power. In this great spirit of God we are all swimming like a lot of little fish. We realize this as we grow in consciousness. We are all equal, for the same spirit supports the man at the bottom and the man at the top. It is not a respecter of what it supports, and cares not who partakes of it.

In this invisible ocean, as in the visible ocean, we have all kinds of forms.

Each form is equally favored, whether it be a rattle snake or a human being, and yet we condemn some forms as being less worthy than others, and we condemn our fellow men, also. We are all the body of God and living in his power and held fast in it, so we cannot run away, and it is from the

highest to the lowest.

Did we take time to study all the forms in the visible and the invisible ocean, we would know God just as the surgeon knows every part of the human body, and [6] this is what we all have to learn. We are all within him, and can swim from his toes to the top of his head, metaphorically speaking, once we learn.

What then is the difference between the natural man and the so called divine man? The difference is in vibrations. Here is the half-way point of the ocean. Going up, the vibration appears to have less pressure and to be finer. Going down, the vibration is heavier and coarser. Finer vibrations are brighter and better because the pressure is less heavy. At the bottom, things are coarse and heavy because the pressure is strong. Suppose we divide the human body as we have divided the ocean, the same principle holds true. The dividing point is the waistline. Below the waistline, man vibrates toward the earth plane, and above the waistline, towards Spirit or God. And yet it is all one, for all is God. The man whose thoughts dwell in the lower nature vibrating on the lower half, is in the animal passion state; vibrating on the higher half is the Christ passion state. We must all destroy the dividing line, and know only one state. The way to destroy the line is to permit the master mind, or the higher mind, to go all the way down and up the body. It then brings the animal passion to the heart and control takes place at that point. When man reaches this understanding, he will also understand his fellow man, and how to serve them.

He will understand that his fellow man is like himself and that the same power supports all. He will understand what mastery is and how to become a master. "Know thyself, and you will know all things." Once we get into the master vibration, we shall be able to go from [7] the bottom to the top, know all there is to know and have total dominion of the Father's house, which house is the invisible and visible ocean.

That substance which the visible ocean feeds its children is called Divine Substance, and no form can exist without it. Then what name shall we give to this invisible ocean, if not the heavens? We think of heaven as the unseen world. This invisible ocean is the invisible kingdom of God, or the spiritual world. Where does it manifest or start? Just as the material ocean starts from the floor of the ocean, so does the invisible ocean start from the floor of the universe, the earth.

No one can measure its depth or height. We walk in the invisible ocean, and we call it space. There are 15 lbs. of pressure per square inch against our bodies from the atmosphere. Walking, in space, we are as much in heaven now as we ever shall be. There is no other heaven than this invisible ocean in which we all live and move and have our being. As you become conscious of the finer layers, you advance in this heaven, just as the master fish saw the light of the sun, so the master among us sees the Father's throne and is conscious of the tremendous light from which the power flows.

People speak of earth-bound spirits. How can they be earth-bound when both heaven and earth are one? The earth is only the floor of the heaven, end even we who are supposed

to be earth-bound now, as we walk, on it, our feet are touching the earth, but our heads are in heaven, immersed in so-called space. [8]

Jesus understood the principle when he said, "The kingdom of heaven is at hand." It is right here and now, if we can only be conscious of it. Our bodies do not stand in the way of mastery. Jesus transformed and dissolved his body as it was necessary, and could then use it for all purposes. The principle is the same when we go traveling. When going to Alaska in the winter, we prepare for it with proper clothing, and again when coming back. Jesus could use his body to go to the bottom or to the top of the Father's house. Why should this appear strange to you? Does not a drop of water evaporate and go back to the gaseous state? The same water in the gaseous state solidifies and comes back to us again as water.

The law is perfect. It remains for us to see how the law works. Knowing the law, we can do anything under the sun, no matter what it is. One night, I went to a circle where fourteen people were gathered, and they will all support what I am saying now. I actually brought forth water from out of the atmosphere. We must have harmony, however.

From where do you think the apple gets its color? Or the peach, etc.? God made everything with the same consciousness that is in you. The invisible ocean furnishes everything, everything is in it. All you have to do is to unite yourself with the total, know that it is there and you will have it. Then you may have it at your command by the slightest

word. But it must be of absolute certainty without the slightest doubt. When you get to that point, you will have what you want. How many [9] people have come to me and have told me that they put this principle into operation and got what they wanted! But let us not sit under the tree and wait for the apples to fall. God helps those who help themselves. We must make an effort of some kind.

This invisible ocean of ours is so great that we are like grains of sand in it. It has no limits. No matter how high man goes into consciousness, it will still have no limits for him. No man can measure its power. Everything is right here. We can sit here now and be wherever we want to be.

I can sit down in a restaurant and begin to think, and beautiful visions will come before me. I am conscious of the restaurant and at the same time I am a great distance away. I go where I like to, and see what I want to, and who I want to. One does not have to go to sleep to travel like that. One can be about the Father's business in the invisible and visible at the same time, as the mind makes it possible to play the piano and answer important questions at the same time. In the restaurant, I am feeding my body while I am a distance away. I am conscious of where my body is, and of where I am. Anyone can do this.

The principle is simple. If a cannon is fired in the middle of the ocean, the total circumference feels it. It will vibrate down and up sending the message through. Wherever vibrations may be set off in India, Tibet or on the throne of heavens, I shall partake of the vibrations if I understand

where I am. And I am in the invisible ocean [10] of vibrations or consciousness. Let us all have the urge, that great desire to Learn and understand these vibrations, or consciousness, to know them for what they are. We shall then comply with the laws governing them and go on to that great destiny. [11]

Many people today are wondering what is meant by the Fourth Dimension. In reality, there is no such thing. As long as you have dimensions, you cannot have oneness. Dimensions are nothing but divisions. There are no divisions in God's mind. What is called the fourth dimension is what man is becoming conscious of today. We have seen the three dimensional world in which we live. Now people are beginning to get into the consciousness of the invisible world from which all things come forth. They are acquainting themselves with the ALL instead of the parts.

The fourth dimension is responsible for all form. It is a united condition of all elements, as we united earth, fire, air and water. The earth is visible, the fire is visible and the water is visible. The fourth element is the air, which is invisible. Yet air has always been here, and has been interblending with the other three dimensions because they could not have existed without air. That had to impregnate itself into the other three in order to support them.

The breath of air is supporting the three dimensions, we only see the three dimensions, but do not see what supports it. When we study the air, we see that nothing could exist without it. We then see that the invisible which we call air is supporting the three coarser elements. There is no division between the visible and the invisible. It is all one.

The earth, fire and water could not bring forth a manifestation without the aid of the atmospheric air. The fourth dimension coming in, which is the fourth bar, makes a complete square condition, and then a manifesta- [12] tion of

form takes place. The manifestation is the fifth.

There is no division anywhere. Fire is in the earth; water is in the earth; all interblending, the atmospheric air is right on the surface, interpenetrating into the inner layers, and also interblending.

When you realize these things, you begin to see the invisible or real kingdom of the universe, and then you call it the fourth dimension, but it is not new, and it is not a dimension, for it takes them all to bring forth a manifestation. So they are all one.

The mind acts in the same way. You have the conscious, the subconscious, and the superconscious, the three dimensions in which you have been living all the time.

Someone may have told you that you acquire the superconscious through study. This is not true since you have been living in a three dimensional world and therefore have always functioned in the three levels of mind, but through study and understanding we learn to bring the superconsciousness into harmonious contact with the conscious mind, thereby clearing the way for the Expression of Divine Mind. But the Superconscious Mind is not the Divine Mind.

There are criminals today, who are superconscious. They can feel the slightest move of someone when they are in danger. That is supersensitiveness. You will find that some safe blowers are so sensitive that their fingers can feel the slightest click of a dial. That is supersensitiveness, which is superconsciousness, the negative end of the same force.

In order to be sensitive you must be conscious. So [13] supersensitiveness is superconsciousness. Supersensitiviness is

a state of feeling, and a state of feeling is a state of consciousness.

You can have supersensitive feeling, hearing, sight, touch and smell. Each one carries within itself the triangle because each one can function in its own realm, but you have not contacted Divine Mind with all of that.

When you contact Divine Mind you no longer look to death as an end, for Divine Mind would not let you drop your body, so you would have everlasting life in your body once you contacted Divine Mind. Neither will you condemn anybody when that state is attained.

Teachers have been fooling you if they made you believe that they have contacted Divine Mind. They have fooled you in the same way about the Fourth Dimension. There are no dimensions for the Father is the total, and the Father cannot be divided against Himself.

You have been expressing the three types of consciousness up to the present day without knowing anything about occult laws. Now what you are trying to learn is through the three states of consciousness, subconsciousness and superconsciousness to attain to the Divine Mind that no man on this earth plane since Jesus has attained.

When I get into the consciousness of the Masters, I get an idea of Divine Mind, but if I contacted it now, I should vanish in the twinkling of an eye. If you were unprepared for it, you would be burnt up.

God is a consuming fire.

Even when you are talking inspirationally, you are as far from Divine Mind as you are from China. All you [14] have

actually done is permitted a mediator to come through, which is the Christ truth, to permit the flow to take place FROM the Divine Mind by reduction to fit you in your present state of development. When you contact Divine Mind you will be like a glowing light. That is, the fourth and will connect the other three to make a complete square. Out of that comes the center manifestation which is the Father.

We are on the road to the so called Fourth Dimension. When the air and the other three dimensions are united, a form comes forth. The breath of life draws them up. That is why the plant comes to the surface. The invisible element pulls it up.

You are now trying to sensitize your consciousness, that is all. When you get so sensitive that you can contact Divine Mind, you will be one with the Father.

Why did God appear in a cloud? He was so powerful that He had to insulate Himself. Were you to contact a fine electrical wire that was not insulated, you would be killed and yet that power is as nothing to the Other.

So you see that we are merely improving on our instruments, our bodies, making them more and more sensitive, so that, in time, through constant improvement, we can contact Divine Mind without being burnt up.

Here is a piece of wood. I burn it. It has been transformed into a finer state now, and fire has no more effect on it. The piece of wood was in a coarse state and the ashes are in a fine state. Your body is like a piece of wood, and it is burning up gradually. [15]

THE CHARM OF LETTING GO

Jesus made a statement, "I am in the world but not of it."*
What did He mean by that? There are several meanings. I shall
give you one.

Suppose I have beautiful clothes which I love. Should a fire
destroy them, I shall still be a man because I never was a slave
to my clothes. Should I be enslaved to them, I shall go almost
insane when they are destroyed, because my master has left
me. Instead of having been the master of these clothes, the
clothes were mastering me.

Jesus did not permit the many things of the world to master
Him, but He made use of them. He enslaved the things for
the good that He could get out of them, but He never allowed
them to enslave Him.

How many of us can let go of things that are not serving us
well? How many can forget yesterday and live only in today? I
doubt whether one in a million can do that today.

When things have served their purpose, they have gone
never to return. Yet we allow them to return as a memory to
torture us. The person who can forget is the person who is
forgiven. The person who cannot forget is not forgiven.

People who go to confession go with the idea of forgiveness
for the things they have done wrong, but the very day they
recall conditions of yesterday. They then go and do the very
things that they confessed were wrong. Why? Because they

---

* John 17:13-14, "And now come I to thee; and these things I speak in the world,
that they might have my joy fulfilled in themselves. I have given them thy word;
and the world hath hated them, because they are not of the world, even as I am
not of the world."

lack assurance within themselves that they are masters of conditions. They believe that conditions can master them. As long as conditions master them, [16] they must serve conditions, and they allow conditions that existed long ago to torment them. When people are masters, they are greater than conditions ans so can dissolve them and forget.

Suppose man had not evolved into a higher intelligence than the cave man? He would still be mastered by those caveman conditions. To evolve, he had to forget the caveman ideas in order to bring forth better ideas that could serve him better. He had to forget.

Man progresses in life in just the same way. He has to forget the old ideas, the old laws and the old customs, and has to bring in new ideas and new customs.

How can you bring health into your body, if you cannot forget sickness? How can you bring forth plenty, if you cannot forget an empty pocketbook? How can you bring forth mastery, if you cannot forget weaknesses? It can not be done. You cannot serve two masters at the same time. You must forget one and serve the other. If you cannot forget, you are serving the one you can not forget and you are ignoring the one you should be serving.

The oak tree is the giant king of the forest. He laughs at the willow tree below him. A heavy storm comes along. As the oak resists the storm, down he goes. The little willow, during the same storm, sways its little head down to the ground. When the storm goes away, the little willow smiles again while the oak is down.

The little willow obeyed conditions for the time being. It complied and cooperated with the storm. The oak thought it could withstand the storm and fell. The same [17] is true with us. We believe we can stand up under any pressure. We stand with a crystallized mind expecting to go through life that way. It cannot be done. We must be able to sway when a heavy storm comes so that we may not be injured.

Let us take a few instances of crystalization of mind. You have the religious person who has placed himself in the kernel of one creed, and he believes that one is the best because his grandfather believed in it. A tremendous storm comes along in the form of a movement that is trying to abolish that creed, and teach joy and happiness to man in this world.

The man mentioned above gets stirred up and tries to resist it. The shock may be so great to him that it will kill him. That is what happened to William J. Bryan at the Evolution trial.* It says in the Bible that "God will destroy the one that tries to block His works, and He will glorify the works and the one that defends them." Bryan stood crystalized like a giant oak. He was looked upon as such by some people. A little man whom the public did not know very well appeared in defence of the theory. Without a warning, W. J. Bryan vanished. It is said that he overloaded his stomach, but can it not be true that he antagonized himself so much that his food so poisoned him? He could not stand the pounding against him by the Defender. He could not sway in the balance. He swayed finally to the limit and went down.

I know a man in this city who was called John. He was

___

* William Jennings Bryan (1860-1925) was a politician from Nebraska who died shortly after gaining national attention for attacking the teaching of evolution in the 'Scopes Monkey Trial' of July 1925, in which school teacher John Scopes stood accused of unlawfully teaching human evolution in a state funded school.

a fine old man. No one, however, could change his belief. A man made a bet with him that a certain condi- [18] tion could be brought about. The man won. The old man died of heart failure.

If we allow ourselves to get crystalized, we all get hurt. We can find fine examples of this law in mechanics. If a man does not look after his automobile, parts of it will crystalize. He may even get killed through his own carelessness when they break.

We must be able to forget before we can get. We must let go of the old in order to get the new. Can you place new furniture in a room unless you remove the old? Let go of the old and take what is coming now. How many people have stuck to jobs for a life time when they could have bettered themselves had they had nerve enough to leave?

How many people have failed when certain conditions came up because they were afraid to let go of something fearing failure if they did? And so they did fail in life. How many have made themselves unhappy, because they could not forget what someone said. Not being able to give up the old for the new brought discontent and heartaches.

How many disagreeable conditions come into our lives because we could not give up the thing that has served for the new thing that wants to serve US? I have been able to keep as young as I look by being willing to let go of the things that were no longer serving me. Why should I carry bones in my pocket from which the meat was taken off a long time ago? I want to keep my pockets free and empty for new things that will serve me. Never mind yesterday. The door must be open

so that greater things can come tomorrow. Do not lock the door of your mind. [19]

You shall materialize anything you want, when you have learned to let go of all conditions that hold you back. I cured a lady of a paralytic condition by giving up the idea she was paralyzed, altho the fact faced me.

The trouble with development in the occult is that people will not give up, and they try to dominate too many conditions at one time. You must take up one condition at a time. Let one condition at a time go out of your existence. It will never bother you again. Learn that rule and you will never be hurt. If you go to a doctor, feeling ill, and he tells you that you are not ill, should you insist that you are ill you will be. That is because you are not able to forget and so the doctor could not help you.

We must forget ourselves as we are at the present time, and bring in that real knowing of self which has not as yet come into manifestation because we judge by form. A man is not a man by the way his form appears. Man must forget form before he can be one with God. He must even forget that his body is not he. He must forget all that, and just as soon as he does, he will be the real One manifesting and he is then ready to take on the Godhood condition. "When the son of man is lifted up, then he shall know that he is I."* "When you see me, you see the Father."† But you cannot see the Father by seeing Jesus or any other man, if you are judging by the framework in which the son is living. As long as you see the building, you will never see the man inhabiting it. Ignore the building and you will see the man. The building is being cared for by the one

---

* John 8:28, "When ye have lifted up the Son of man, then shall ye know that I am he, and that I do nothing of myself; but as my Father hath taught me, I speak these things."

† John 14:9, "He that hath seen me hath seen the Father; and how sayest thou then, Shew us the Father?"

in the building. [20]

When a man begins to see the real man within the building, he begins to climb. He begins to be a real man. But how many want to take on the new man? How many want to forget the old conditions? How many want to keep silent concerning what they consider wrong? How many want to forget slights, but hold them in their hearts? God does not hold any grudge. When you contact Divine Mind, you will not say no word against another man.

The masters in Tibet are not worried by anything. Nothing bothers them for they know what is theirs they will have. When a thing is not for them, the one to whom it belongs will have it regardless of how they try to hold it. What is mine is mine forever. I am willing to let go of anything that does not belong to me, if the time has come for it to go. If I cannot trust you, I cannot trust myself. What is in me will reflect on you. When I stand before you, I reflect upon you that which I am, and you will reflect that towards me. Were you all alone in the jungle, could you condemn any man for anything? There would be no man on whom you could reflect yourself. You must feel that condemnation and it must be in you. A man who kills has murder in him. Those who want to kill a murderer are murderers at heart, otherwise we should not want to kill him. A child cannot say "A" until it learns the letter "A" and reflects the letter "A". What I reflect in the lower nature is because of the second or lower spirit of man from the waist down. What reflects from the waist up is the higher spirit. We can balance the two when we know how. [21]

Never be small enough to knock. Allow yourself to condemn yourself in order to see your better self. When you can do that you have won a great battle in life. Be ashamed of yourself when you condemn. You will hate to see yourself in the mirror when you do that. The small self is the Intellect. The big self is Divinity.

Let go of the little self. Hang on to the big one that moves the universe and your way will be clear. Let's be big enough. How many actually are? We must be before we find ourselves.

Every year a man takes an inventory of his business. Are we not a business house within ourselves? Don't we execute many conditions? Let us find out what we have done and what we are planning to do. When a business goes wrong, we get an expert to check up. Let us do the same thing. Let us check up on ourselves and see what is wrong. Then alter the wrong condition. We must educate ourselves as we educate a child by steadily pounding at it.

And here I want to say, do not tire people by thrusting something down their throat. Let us wait until they are ready and then give it to them. The soil must be ready. Let us learn that law. I must water that soil, get it into shape, and then plant the seed. When I plant the wrong seed, a lot of weed comes up. What must I do? What I sow I reap. Before I can plant new seed, I must reap the weeds. I must apply wisdom and I must sow the proper seed. Be sure you are planting the proper seed so that weeds will not come up. Plant the seed in the heart of sincere seekers who are trying to understand themselves [22] in order to unite with

the Father so that the Father may express, so that the son of man shall be lifted up. That is all we have to do.

I expect to live 150 or 200 years, but I cannot say that I shall be allowed to do that. I may have to pay a penalty for having sown weeds, and then I shall have to come back in order to pay for what I have done.

In Divine Mind you are everything and have no worries, but remember, we can not wear more than one suit of clothes at a time. We can eat only one meal at a time. We can have only one thing at a time. When that is gone, we get another. When you have something that does not belong to you, it is taken away, but nobody can take anything that belongs to you.

Be willing to let go of everything and anything when the time comes for that to go, and for the new to come. Then you will enjoy life.

Humanity has long been waiting for this great move, and today we are gathering together those who realize that the time is at hand for the co-operation of souls seeking the highest, and to be bound together only by the ties of the Infinite Love and Wisdom. Aspiring Allsouls to that Allgood and true! The Royal Order of Tibet is for the purpose of establishing the All into One Eternal Life Progress. Its aim is to bring about the best possible conditions for every soul. It presents All inclusive Cosmic Teaching, a scientific and practical Master course of the Universal Message, and a method whereby the individual attains access to mystery in a simplified way. [23]

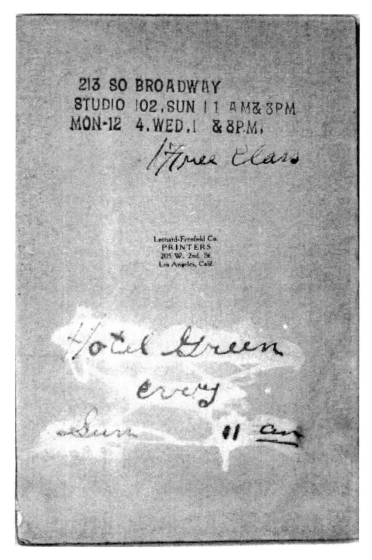

End page with addresses and class times, with a handwritten note offering "1 Free Class".

TWO ARTICLES

The two articles by George Adamski in the following section have not been published before. It is unknown when they were written, but they were distributed among members of the international Get Acquainted Program, probably in the early 1960s, and tie in seamlessly with many of the chapters in his book *Cosmic Philosophy* (1961).

These texts are included here because they are of particular interest in conjunction with those in his first publication, *The Invisible Ocean*, as they elaborate on the same themes of universal and individual consciousness, and self-mastery.

The first, 'Transformation of Body Consciousness', discusses the need to shift our identification from the lower, personality self that we see in the mirror, to the Higher Self, the soul. In the second, 'Individual Analysis and Thought Control', the author elaborates on the need for detachment from the temporary manifestations of consciousness, from the not-Self, in order to identify with what is Real and eternal.

TRANSFORMATION OF BODY CONSCIOUSNESS

Part I.

We can bring ourselves into a higher state of consciousness through various channels. As we transform thought, we also transform the body, because thought has a dominion over the body. As it is in the mind, so it is in the body, and nothing but thought will transform the body. The thoughts which we entertain within ourselves draw like conditions from the universe unto us. If we wish to rise in consciousness, we must forget all of the past conditions which have already served us.

We must rise into that higher understanding of a limitless being. The higher thoughts will eliminate all wrong conditions within the body. The higher thoughts dwell in eternal peace.

This peace comes as we begin to understand ourselves. It cannot come in any other way. Before we can attain harmony we must understand ourselves. Our consciousness is like a huge transformer and we can take in as much power as we wish and transform it from one part of the body to another.

We can control the body completely by our consciousness if we are conscious of conscious power in thought. When our consciousness is not functioning properly, we may try every cure and accomplish nothing. We can see, therefore, how necessary it is for us to understand the conscious power of thought. When our consciousness is in complete unity with the Father's consciousness it is limitless for we are His counterpart. Knowing what we are we must hold fast to that which we want and eliminate thoughts which we do not want. We are then bound to get results if that which we want is right for us at the time. Otherwise we shall get what we want at the proper time. But man must have faith and confidence in the workings of the eternal law.

We can use this as an example. Suppose a man wants to change a condition into a higher or better condition then he must first change his thoughts into a higher state and must have faith that the better condition will follow. If he has any doubt in his mortal consciousness he will block the condition from appearing. A doubt as small as a mustard seed will keep it from coming through, but should he have faith as large as a mustard seed, without any doubt, he shall have the attainment of any condition.

A man must have certainty in his conscious thought that he can and will rise and then the higher conditions will take place. By lacking certainty he lowers his mortal consciousness instead of raising it. Man has risen from the savage state to the present civilization only by wanting higher and better things but he had to be absolutely certain that he could have them and that certainty led him to success.

Yet the man in a simple state understood this power of conscious thought in his body better than the civilized man of today, for the simple man trusted in Divine Providence, perhaps to an extreme and that is why he hurt himself at times, for the Father does not want us to be extremists. The present day man takes advantage of everything and everyone because he has no faith in the Father of Creation. The simple man did not store up any more than he could use over a period of time, while the present day man stores more than he can use and cares nothing about his fellow-man's needs. This condition has come about because of the transformation of man's mortal mind, from faith to no faith. The simple man may not have had the knowledge that we have today but he was more just. The simple man could control his thought while modern man has acquired a powerful mortal consciousness that he is not capable of controlling. He has therefore lowered his character.

Instead of mastering his thoughts as he should he allows his thoughts to master him and he is beginning to realize that he does not know how to master the power of conscious thought.

His thoughts run away with him, confusing him and he lives in a state of fear instead of in a state of certainty. Man today fears everybody for he knows that he is powerful enough to harm and he fears that his fellow man will use that same power to harm him. Had he taken time while he was transforming his mortal consciousness by analyzing every step that he took, he would have his thoughts and everything else about him under control today.

That is how man's dominion grows. His power would not frighten him as it does for he would know the laws of thought and would know how to use it consciously and properly. We have a similar principle in electricity. Electricity will kill if it is mishandled and will serve if properly handled. The same is true of conscious thought, it is powerful enough to kill. Doctors call that condition a shock.

When we know how to use this power it is then our servant. The wrong use of thought destroys us but the right use brings a harmonious condition, for lasting joy comes only when we learn to be the master of our mortal consciousness. We have been given a free will for one purpose and that is to master and have dominion by submitting our will to the higher will within to guide us. Our Father never intended man to be dominated by anyone or anything. He gave us power and intended for us to know how to use it. When the so-called mind is under proper control it does the proper thing at the proper time in the proper place and works for the good of mankind. This is the higher will expressing. This conscious awareness must hold the reins and guide the mortal consciousness instead of being driven by it for then fear will disappear.

A man in a dark alley is afraid and watches his steps very carefully because he cannot see what is ahead of him. The carnal mind is working in the dark and is constantly afraid of the steps to be taken fearing that something will arise to block or injure it from going on. The man functioning in the carnal consciousness sees no light ahead of him. The fear that he projects causes the animal to attack him. Man has thrown away the Divine reins by which he should be guided and now he suffers much sorrow and pain but finally he awakens to the fact that something is wrong and begins to look into the universe for help. He then turns to God for understanding. He is the prodigal son going back to his Father. He thought he could go wildly into the world and have his own way but finally realized that he could not control the conditions of life by his mortal will.

Man can partake of the overall intelligence when he is willing to take the step forward and can get what he really wants. When he takes this step he finds that he has dominion over the lower elements through his feeling of unity and he serves the total rather than the personal desire. Without any true knowledge man took conditions upon himself which were never intended for him.

The overall intelligence never assumes anything because it knows everything and is unlimited. It is the personal consciousness that gets man into trouble for it has assumed the wrong condition. When man is hurt enough he begins to look to the Father, or the reality of life. It sometimes takes all of the strength that he has to eliminate the wrong assumptions when he finds himself in such trouble.

How does man find himself? When he realizes the reality of life, when he realizes his true self, he then must step into a new garment and learn to say "Thy Will Be Done". When he does, the wrong conditions will dissolve and the sorrows will go with them.

When we unite with this knowing we must throw away the old thoughts and take on the new attitude towards life. In other words we must wash away the wrong ideas but the carnal consciousness cannot do this work by itself any more than a little child can wash itself. The mother must cleanse the child and so our impersonal consciousness is the mother that must cleanse the personal mind. That is man's only salvation.

Part II.

Our intellectual or carnal consciousness, though lacking in understanding, has great power, and because it lacks this understanding it is a strong factor in destruction. Working in darkness, it is continually investigating and getting into those wrong assumptions.

When we submit ourselves to the higher, or Cause–consciousness, at first we are guided much as a child [is] guided as it grows up. Then we gradually acquire more and more strength and learn to use the strength for the better way of life. The boy who wants to be an electrician must have a master electrician to guide him along his path of study and the teacher tells him what to do.

The boy is then promoted as he learns. He may get a shock here and there if he is too inquisitive and tries for himself, but he will not get as hurt as he would, if he did not have the training as he learned the laws of electricity.

The electrician guided him along the avenue of knowledge in controlling the power which we call electricity. He has been guided step by step from a state of no control to a full control of this power and so the boy grows stronger and wiser and begins to apply his knowledge with wisdom.

The power then cannot hurt him for it serves him, and he can use the power at any time in any place.

In order for us to transform ourselves to have the power to overcome our present state of conditions, we must submit our lives to the great Engineer of the Universe, or the overall consciousness. Then the Cause–consciousness will explain all things to us. If we pay attention and heed we shall grow into an expanding understanding and use this power for the good of everyone. We shall learn how to control our intellectual thoughts so that they will be our servants. Knowledge is the path to the control of our powers just as the boy electrician had to learn and understand all the laws of electricity in order to learn its control.

God has given us a way to attain everything but we must know how it can be done. It is high time that we return back to the Father. Man is running wild with the power that he has. Not understanding it, he runs into wars and all manner of destruction. Cause-consciousness is the source that knows how to meet all problems and how to adjust conditions in order to bring about peace on earth.

The only hope for mankind is to transform his consciousness and be born again. We must use our will power to accomplish this for it was given to us for this purpose only, to compel our intellectual consciousness to be guided by the "Thy Will" from now on, and then our lives will begin to reflect its power. How? Does not a pain in your body reflect on your face? And by the same token the good that we feel reflects in our lives.

By permitting the higher consciousness to guide, you will be able to control conditions for now you will be controlling that power which you had been using but in a negative way. As you come into understanding you will use the power in a positive way, and you will build up good conditions. Look at the people in the streets, uncertain, wondering, not knowing which way to turn. It is because they are not controlling thought but allowing thoughts to control them.

By allowing the higher thoughts to control, you will be led to the exact contacts and a realization that everything is well, and as it should be, including the purpose for everything. This awareness will place a smile on your face and give you strength or vitality so that you can go on eternally without tiring.

But you must use your will power to compel yourself to function in this consciousness. In the olden days there were wise mothers who knew how to apply the law. Not having the means to give their boys and girls an education, they would take them to see the statue of some great man or woman and tell them all about this person. They made the children feel that they were like that person and could do the things that he or she had done. The children began to take on that consciousness step by step until their feeling became so strong that they wanted to be like that great man or woman. Gradually they became conscious of the power within them and felt they could succeed. We can transform our consciousness not by saying but by feeling.

We must feel certain that our actions are right. We must feel the joy which leads to an unselfish expression for we grow through feeling, which is consciousness, when feelings are controlled. We will then become what we feel for the self is transformed by conscious thought.

Suppose you were to go before some prominent man. If you felt that you were nothing and he a great man you would wither away, but if you felt that you were equal to him you could meet him face to face and feel perfectly at ease. We should all feel that way. Is not the President a human being like us? His position means nothing when it is compared with life eternal. If you feel like a great man you are working in the same vibration as he is. Then he will more likely grant you the things you want.

If you are looking for work do not think that the employer is superior to you. If he feels that you think yourself inferior, he automatically becomes filled with vanity and tries to take advantage of you. Be on his level and he will treat you right. Never lower yourself to anyone. You are as much a child of the Father as the next fellow and God is not a respecter of persons. We must learn that only through feeling can we make ourselves into a bigger man or woman. Rise before God and stand before Him. Do not shrink.

Jesus said it is not robbery to be equal with our Creator when we feel the consciousness He has granted us. Then surely it is not robbery to feel equal to your fellow man when the consciousness is transformed into the higher self. Then you will gradually eliminate pains, aches, sorrows, and all negative conditions.

It is the higher awareness that teaches us that we are the sons and daughters of the Father and that our power is unlimited since His power is unlimited. Through this awareness we learn to be happy for we know that the Father always provides. We live in certainty and never worry, for it is only the carnal consciousness that worries since it knows no better.

The man who dwells in the Universal thought is always certain and always joyous, vibrating that joyousness to everyone who comes in contact with him. He knows that all of his questions will be answered and that he is being guided along the pathway of life. He knows that all power is for him to use [and] that the universe wants him to use it.

We must all at some time learn the laws of the universe but because we have been hurt we have tried to run away from them. Yet when we learn to work with the laws we let our free will be guided by the Thy-Will and we shall then begin to transform our whole being. We shall find heaven right here on earth, and now.

INDIVIDUAL ANALYSIS
AND THOUGHT CONTROL

Almost every thoughtful person finds himself at times pondering over the various problems which face him in the field of human evolvement, and from time to time glimpses a little light here and there, yet not enough to connect or make a complete picture of his relationship to nature and mankind.

The human mind is restless and jumps at conclusions thus it does not see the Real back of the effects. It does not remain quiet long enough to experience any peace for itself and therefore is unable to make full connection with Cosmic Consciousness.

The most important step that any person can take is the practice of self-control, especially in the field of mental action. Thoughts pass through the mind with great rapidity and if they are not controlled they may be likened to a run-away train whose speed cannot ever compare with the speed of thought, but when controlled is very liable to cause much danger to itself and other forms. The engineer must have full control of the machinery which guides his locomotive or many lives would be endangered and he would be responsible for all of them. Were he to act in a restless impatient manner as the mortal mind acts he would not be alerted to the importance of bringing the train under control.

When a human being does not control his thoughts he does not realize the danger his quick acting run-away mind might cause to others. A thought maintained by anyone is projected into space; it cannot be withheld, for action is continually going on and must take effect somewhere. The thought affects the other waves over which it passes and also lodges within the form through which it is expressed.

If it is a good thought it will do much beneficial work in space as well as within the body of itself setting up harmonious vibrations in all that it contacts. If the thought is not [a] pleasant one the same action will take place but it will do great harm in space where it will draw unto itself conditions of a like nature and cause an inharmonious state within the body, confusing all of the peaceful cells. A thought continued in such a channel will eventually unbalance the whole body and produce pain within it.

A balanced body is one in a natural state, peaceful, harmonious, thus allowing all of the cells to do their work in a normal way. In this state the body becomes lighter and attuned to Cosmic Consciousness.

Everyone wishing to know mental and physical well-being must learn to control his thoughts. He must realize that thoughts are powerful and can be directed in either a constructive or destructive channel according to the choice of the individual. Habits are sometimes very difficult to break and a selfish thought that has been allowed to run wild for some time will be harder to control than the destructive thought which is transmitted immediately.

It is strange that this being called man, who has free will will use it in the channel which is most painful to himself. Sometimes this is done intentionally to gain sympathy from others. Sometimes he does this to attract attention to himself for that comforts his ego with the idea that he is the outstanding person in a group. He fails to take into consideration that others are not attracted by that kind of effect, for since consciousness is not a respecter anyone can see through the effects to the thought [behind] it.

This, of course, disturbs the personal ego of the individual and after failing to make an impression is hurt and tries many schemes to re-establish itself in the good graces of those it wishes to impress. What an enormous amount of energy is wasted through actions that the mortal projects in order to convince itself that it is all-important.

It was never meant for man to use the power and strength of his being in the channel of vanity and selfishness. Man was meant to be a humble servant unto his greater self and until he realizes this fact and puts aside his stubborn dominance and self-exaltation he will be unable to make the right connection with Cosmic Consciousness which is the Creator of all that exists.

Laying aside these falsities of itself is not an easy task for they have been given total recognition for such a long time that they have become the controller of man's being; however, if the student is to find his place in life he must transmute his selfishness into higher fields of action.

This could be done most easily, perhaps, by finding channels of service which would be beneficial to many instead of the few. The more one enters into the service of others the less he thinks about himself as a personality. When man learns to look upon himself as a channel privileged to perform a duty regardless of what it may be, seeing only that it needs to be done and happy for the opportunity to serve, he will then be about the Father's business. And who is to say what that business should be or who is to take charge of each of its different branches. One man may find his greatest work in some obscure place where he may never be known as a personality but so long as each one is acting impersonally upon the thing that presents itself to be done he is serving well.

The purpose of life is not personal attainment but the unification of all action in the cosmic sense. Any thought that helps keep an individual thinking in this vast field should be maintained at all times. If the mortal mind holds the cosmic thought for a time and then allows the thoughts to slip back to self it is expending, uselessly, much energy that could be turned in a constructive channel of service.

Each person must find his own way for elevating his mortal thoughts. He must study himself and the way his mind functions. He must watch his own actions and reactions and find which thoughts will produce the most peaceful and harmonious conditions within his own body. There are no two individuals that are exactly the same. A teacher may give the universal laws to all of his students alike but each one will apply the laws in a little different way than another. Each man has a destiny to work out in each life – to one it may be the overcoming of fear; to another it may be a tendency towards jealousy which must be eliminated. One may gain his rewards through hard work while another may gain it through deep devotion.

A man of great will must learn humility while an individual who is naturally humble must find the way to unite that meekness with positive action. The path is a little different for each individual. Man himself must find himself, his duty as a form, his relationship to all forms and his oneness with the consciousness and intelligence which is manifesting through all.

The vastness of knowledge and wisdom which can be attained by sincerity, honesty and love is unlimited, and the amount of service an individual can give is also unlimited.

What a privilege man has in being allowed to will himself to such a life of useful service. It is well worth any effort that a man may make to steer his ship of life on the true course that leads to a vast concept of the interrelationship of all life. Then he can truthfully say as Jesus said, "I and the Father are one."*

*John 10:30

THE ROYAL ORDER OF TIBET

1932-1940

Clippings

Humanity has long been waiting for this great move, and today we are gathering together those who realize that the time is at hand for the co-operation of souls seeking the highest, and to be bound together only by the ties of the Infinite Love and Wisdom. Aspiring Allsouls to that Allgood and true! The Royal Order of Tibet is for the purpose of establishing the All into One Eternal Life Progress. Its aim is to bring about the best possible conditions for every soul. It presents All inclusive Cosmic Teaching, scientific and practical Master course of the Universal Message, and a method whereby the individual attains access to mystery in a simplified way.

Above: Closing paragraph of *The Invisible Ocean* (p.23), where the name of The Royal Order of Tibet appears in print for the first time.

On the end page of our copy of *The Invisible Ocean* (see p.24) a stamp indicates that meetings of The Royal Order of Tibet were held at

Studio 102, 213 South Broadway, Los Angeles (photo left), every Sunday, Monday and Wednesday. Below it is added: "Hotel Green Every Sun 11am". Adamski's calling card (p.41) names it as Hotel Castle Green in Pasadena, where classes were taught until late 1933, when the Temple of Scientific Philosophy was established at Laguna Beach (see following pages).

(Photo of the original building as displayed in the front window of the current building. Google Street View, January 2017.)

(photo: Wikipedia)

A 1923 photograph of Hotel Green, Pasadena, where George Adamski taught classes until late 1933. With the addition of the Central Annex (background) in 1899 the hotel became known as Castle Green.

While visiting Adamski at Palomar Gardens in 1954 a reporter from a local newspaper spoke with one of the attendants, an elderly teacher. She told him she was a regular visitor to the weekend meetings which Adamski held at the Palomar cafe every first and third Sunday of the month: "I have been a follower of his for 22 years..." [i.e. since 1932] Confusing the names of the various aspects of his early efforts, she told the reporter: "He headed the University of Religious Science in Pasadena and Laguna Beach...", adding: "There were about 50 of us." (*Torrance Press*, 7 January 1954)

Prof. G. Adamski

Speaker and Teacher of Universal Law
and the Founder of Universal Progressive
Christianity, Royal Order of Tibet
and the monastery at
Laguna Beach

Headquarters
Hotel Castle Green
99 S. Raymond Ave.

Pasadena
California

41

HEADQUARTERS OF TIBET ORDER AT BEACH SOON

Santa Ana Register,
13 November 1933

LAGUNA BEACH, Nov. 13. — International headquarters of the Royal Order of Tibet will be established in Laguna Beach by Prof. George Adamski, it was announced today by Mrs. Maud Johnson, a leader in the Order of Loving Service, which some time ago established the Little White church in the old Joseph Kleitsch Art gallery.

The beautiful Claude Bronner estate on Manzehita drive has been purchased by the Tibetan order. While the cash consideration was not made public, Bronner invested more than $50,000 in the beautiful home and grounds. Within a year, it is anticipated by Mrs. Johnson, a monastery will be established in the hills back of Laguna Beach that will house about 200 monks of the order. The deal was made through Agnes Yoch West and N. E. West.

Selected students from the United States and other countries will attend a school to be held at headquarters. From there the general correspondence of the order will be sent out.

Professor Adamski, who has established centers of the order in various cities of the United States, first came to Laguna Beach to give a lecture at the Little White church. He informed his congregation recently that the hills back of Laguna Beach were ideal for a monastery and expressed the belief that the money would be found for its establishment.

Professor Adamski will give the last lecture of a series Friday night of this week. He expects to open the International headquarters about the first of the year and at that time will give lectures here. He will then devote his time in Laguna Beach to teaching, for the next year, following which he will return to Tibet for three more years of study.

Mother Lisa Mae Grey, of the Forum of Universal Metaphysics, will continue to lecture Thursday evenings at the Little White church, but Mrs. Johnson, who has been occupying the Kleitsch home, will take up her residence at the Tibetan headquarters, where she will continue her writings.

Floating Power in the Invisible Ocean

he will find fish that will appreciate his knowledge and teaching. As it is with the fish, so it is with us, only that we work on different planes of mental vibrations instead of strata of water. The master among us can go down to the depths of the earth and up to the lights of heaven. Nothing can hold him for he is above any condition. He can descend to the level of people in the coarser vibration in order to teach them about the light. He goes through layer by layer of consciousness, mixing freely with all the people and here and there he finds some who are ready to learn about the light. The same principle works in the visible and in the invisible ocean. We call spirit of God we are all swimming like a lot of little fish the invisible ocean spirit, force, or power. In this great We realize this as we grow in consciousness. We are all equal, for the same spirit supports the man at the bottom and the man at the top. It is not a respecter of what it supports, and cares not who partakes of it.

In this invisible ocean, as in the visible ocean, we have all kinds of forms.

Each form is equally favored, whether it be a rattle snake or a human being, and yet we condem some forms as being less worthy than others, and we condem our fellowmen, also. We are all the body of God and living in his power and held fast in it, so we cannot run away, and it is from the highest to the lowest.

Did we take time to study all the forms in the visible and the invisible ocean, we would know God just as the surgeon knows every part of the human body, and

6

Page 6 from *The Invisible Ocean*, with lines 14 and 15 transposed (see next page).

Transmitted Light

ocean and so I have acted as a sieve and sorted out the light from the dark ocean.

(I will have to explain here that I had handed to the instrument a booklet which had just come to my notice a few weeks previously and had asked him to read it aloud. He had read only a few moments when I noticed that his voice had changed and also the expression of his face. I realized that some one had come in and was doing the reading. He had come to a passage where the lines had been transposed and was having a little difficulty, so I stepped to his side and took the book to examine it and help out in the reading as I had previously read the booklet.

In 1937 the Order of Loving Service, "associated with the Royal Order of Tibet", published a book titled *Transmitted Light*, which was comprised of diary entries and messages channelled by the 'instrument' Latoo and transcribed by the 'recorder' Lalita (Maud Johnson).

In an entry dated 11 April 1933, on page 53, Lalita describes how Latoo was reading from a booklet she had come across a few weeks earlier, when he ran into difficulty with "a passage where the lines had been transposed". This, more likely than not, refers to page 6 in George Adamski's *The Invisible Ocean*, where lines 14 and 15 are transposed (facing page).

This would mean that Mrs Johnson and Adamski became acquainted in 1933, and that she must have been immediately impressed with his teaching and philosophy. Soon after, she offered or agreed to purchase a $50,000 property in Laguna Beach for Adamski's Royal Order of Tibet (see newspaper reports on the following pages).

According to Tony Brunt (in *George Adamski – The Toughest Job in the World*), Adamski's wife Mary was a devout Catholic and very uncomfortable with his 'New Age' teachings. This could be the reason why Mrs Johnson invited Adamski to lecture at the Little White Church where the Order of Loving Service convened, which she had founded in Laguna Beach earlier that year. And perhaps her support also made it possible to temporarily move the Royal Order HQ from Adamski's studio apartment in LA to Hotel Green in Pasadena until the Laguna Beach venue became available.

ROYAL ORDER OF TIBET MAY BUILD MONASTERY IN HILLS THAT OVERLOOK THIS CITY

Buying of Claude Bronner Home for International Headquarters Is Regarded as Preliminary Step to This End

Probability that the Royal Order of Tibet will build a monastery in the hills back of Laguna is seen in the purchase of the Claude D. Bronner home on Manzanita drive, through Prof. George Adamski, who has been lecturing in Laguna at the Little White church, representing the Order of Loving Service. Mrs. M. Lolita Johnson, who has been carrying on the work of the order in the former art gallery of the late Joseph Kleitsch, announces that international headquarters of the order will be opened and maintained at the Bronner home. The final lecture in Prof. Adamski's series at the Little White church will be given Friday night of this week. As soon as the new headquarters are furnished and ready for occupancy, Prof. Adamski will give two more lectures there and then will conduct closed classes. This work may be started around the first of the year.

South Coast News,
17 November 1933

The professor considers the hills back of Laguna an ideal place for the location of a monastery and is said to have indicated that one will be built there at some time in the future.

The Order of Loving Service has a number of followers in Laguna Beach. Mr. Kleitsch, who died last summer, was a member. Mrs. Johnson and Mrs. Marie B. Garth will prepare the Bronner place for use and will temporarily reside there. The price paid for the home has not been made public. The original price of building and grounds is said to have been in excess of $50,000.

(photo: GAsite.org)

This photograph was first reproduced in Lou Zinnstag and Timothy Good's *George Adamski – The Untold Story* (1983) with the caption: "Teaching in the monastery at Laguna Beach, California". However, records show that the venue for the Temple of Scientific Philosophy was built in the 'Mediterranean revival' style (p.50), which is distinctly different from the architectural style as seen in the arches here.

The same photo, but with the Christ statue cropped off, is also included in the Adamski Foundation's expanded 2016 reissue of *Behind the Flying Saucer Mystery*, captioned: "Tuesday Lectures At Castle Green Hotel, Pasadena, Ca. 1935." According to the information in *The Invisible Ocean*, however, the classes at Castle Green were held on Sundays (p.24), while hotel conference rooms are not usually decorated with religious artefacts or images.

Since Adamski was also invited to lecture for Mrs Maud (Lalita) Johnson's Order of Loving Service at the Little White Church in Laguna Beach, this photograph was most likely taken there.

45

(photo: Orange County Archives)

Laguna Beach, California, USA, 1930s

ROYAL ORDER OF TIBET IN ITS LAGUNA HEADQUARTERS

The opening of the Royal Order of Tibet headquarters proved a very happy and stimulating affair. Guests and students began arriving early in the afternoon, bringing with them both food and bedding. They made their own coffee and enjoyed their lunches in the sun parlor, kitchen, dining room or breakfast room, as they wished. Various students had volunteered to make curtains for the large lecture room and some of these were put up at the last minute and many finishing touches had to be made even after some were assembled in the room. It was only by splendid co-operation that all was made ready so quickly, for there was much to be done.

There was music early in the evening and at 8 the services began. Professor Adamski annointed the room with water from a sacred bowl and then gave a much appreciated talk on the subject, "Universal Brotherhood." Many people from Laguna were also present. After the lecture some moving pictures were shown. These had been taken at a previous gathering of students. The rooms were beautiful with flowers and shrubs from the grounds of El Castillo Mio, augmented with palms and flowers from the Virginia Park gardens, which were generously donated for the occasion.

After the services, refreshments were served in the dining room. About midnight the remaining students had another very interesting meeting. And about 1:30 or 2 some very tired people sought the quiet of their own rooms, as far as could be provided, the rest bringing in auto seats and blankets and sleeping on the floor. The next morning some ambitious ones went fishing, while others helped to clean the house and put things in order. During the afternoon the ladies cooked and served a fish dinner, at which 12 were seated, and others sitting here and there with plate dinners. The weather was ideal and altogether the occasion was one which will long be remembered. All realized that this was just a beginning of greater things to come.

There will be a public lecture Friday, January 26, at 8 p. m.

South Coast News,
26 January 1934

47

Tibetan Monastery, First in America, to Shelter Cult Disciples at Laguna Beach

Lamaistic Order to Be Established Here

Purple - Clad Women and Golden - Robed Men Will Study "Ancient Truths" at New International Headquarters

The ten - foot trumpets of far away Lhasa, perched among perpetual snows in the Himalaya Mountains in Tibet, will shortly have their echo on the sedate hills of Southern California's Laguna Beach. Already the Royal Order of Tibet has acquired acreage on the placid hills that bathe their sunkist feet in the purling Pacific and before long, the walls, temples, turrets and dungeons of a Lama monastery will serrate the skyline. It will be the first Tibetan monastery in America and in the course of time, the trained disciples of the cult will filter through its glittering gates to spread "the ancient truths" among all who care to listen.

Inside the great gates, securely sheltered from the madding throng, feminine neophytes in flowing purple will wander through Elysian gardens, seeking to attune the inner being to the practical purposes and demands of a motorized world; men in golden garments with purple collars will endeavour to achieve through logic and science the blissful "mastery of self" which is at least one of the multi - featured goals of the Order of Tibet.

Those familiar with Laguna will instantly recognize the monastery site, for it is the Claude D. Bronner estate on Manzanita Drive. The beautiful dwelling, familiarly known as El Castillo Mio will be occupied by the parent group of the Tibetan Order, while the amphitheater, stage, temple, lecture halls, cell or chamber units and other buildings take form. The estimated cost of the project is $1,500,000, and when completed, it will become the international headquarters of the order. Headquarters are now in London.

Prof. George Adamski

Central figure in the new movement is Prof. George Adamski, sturdy, middle-aged. He is as strange as the cult he sponsors. Now he is an American citizen and served in the World War, but as a child he lived in the ancient monasteries in Tibet and learned the law of the lamas.

His father was Polish, his mother an Egyptian. George Adamski, as first son, was destined to walk in religious lines. He studied them all and very nearly landed in a Catholic monastery but his youthful ideas leaned so strongly to reincarnation that the move was not made.

"I learned great truths up there on 'the roof of the world,' says Adamski, "or rather the trick of applying age-old knowledge to daily life, to cure the body and the mind and to win mastery over self and soul. I do not bring to Laguna the weird rites and bestial superstition in which the old lamaism is steeped but the scientific portions of the religion.

"The Order of Tibet acknowledges God and Christ. We hold to the basic thought of Hinduism, Buddhism, Christianity, to which are added the ancient law of Tibet. But our main object is the application of knowledge, just as Christian Science, Mental Science and other crystallizations of thought are primarily intended to put Christianity into everyday use.

ROBES AND RITUAL

Robes and ritual, Adamski admits, help the novice to set his feet firmly in the path he elects to follow. All churches have found this to be so. A uniform makes the sailor or a different man, and so in the Laguna monastery the robes will be provided.

The symbol of the order is the twenty-four point star. To ancient Tibet it represents the mystics or counselers grouped around God. The women at Laguna will wear a twenty-four-point yellow star and the men a similar pendant of pure white crystal. The decoration will be awarded after a successful three-months' novitiate.

Inmates of the monastery will not be cut off from the world forever. In this respect, the institution will function more like a school. Students can come and go as they choose, though some will undoubtedly make their permanent homes there during the two-year course of instruction that will fit them to go forth as teachers and lecturers. By that time they will have come to understand self-evolvement and the Infinite Will.

Explorers returned from Tibet tell of isolated monasteries in which priests, seeking utter purity and Nirvana in this life, to insure perfect reincarnation, or, better still, direct mingling with the Ultimate without the necessity of more earth-lives, go into cells which are then walled up.

There will be no self-imposed tombs at Laguna, but in addition to the amphitheater, stage and lecture halls, there will be isolated cells or "dark chambers" to which the student can retire to meditate or materialize a dream or ambition. In other words the dark chamber will help carry out the Bible principle, "As you think, so shall it be." When the monastery is fully completed it will accommodate over 200 permanent residents, and many hundreds of " at home " students.

Los Angeles Times, Sunday 8 April 1934

Monastery Of The Royal Order of Tibet

UNIVERSAL

WISDOM

738 Manzanita Drive, Laguna Beach, Calif.

(photo: George M Eberhart)

49

State of California — The Resources Agency
DEPARTMENT OF PARKS AND RECREATION

HISTORIC RESOURCES INVENTORY

Ser. No. 30-2651-17-04a
HABS _____ HAER _____ NR __3__ SHL _____ Loc _____
UTM: A 427960/3712300 B
C _____ D _____

IDENTIFICATION

1. Common name: _____ Anneliese's Preschool - 758 Manzanita _____

2. Historic name: _____ Claude Bronner Home _____

3. Street or rural address: _____ 758 Manzanita _____

   City _____ Laguna Beach _____ Zip _____ 92651 _____ County _____ Orange _____

4. Parcel number: _____

5. Present Owner: _____ Address: _____

   City _____ Zip _____ Ownership is: Public _____ Private __XX__

6. Present Use: _____ Pre-School _____ Original use: _____ Residential _____

DESCRIPTION

7a. Architectural style: _____ Mediterranean Revival _____

7b. Briefly describe the present *physical description* of the site or structure and describe any major alterations from its original condition:

A large Mediterranean Revival house which is elevated with the slope of the lot. Roofline is varied through the use of gabled and shed tiled roofs. The irregular plan gently curves so that the main body forms a slight "C" shape. The house is three stories with two primary wings on either side and a connecting wing in the center with a gabled third story. Entry is at center via a brick and painted tile stairway. Casement windows are used throughout; some have shutters andothers iron gates. The garage is built in at street level and has a tiled hood.

The house sits on a large lot with a great deal of vegetation. The house appears in excellent well-maintained condition with no apparent major alterations.

8. Construction date:
   Estimated _____ Factual __1927__

9. Architect _____ Unknown _____

10. Builder _____ Unknown _____

11. Approx. property size (in feet)
    Frontage __150'__ Depth __140'__
    or approx. acreage _____

12. Date(s) of enclosed photograph(s)
    _____ May 1981 _____

A State of California Resources Agency inventory summary from May 1981 states that the property at 758 Manzanita Drive "appears in excellent well-maintained condition with no apparent major alterations."

In 1971 the property was purchased by Anneliese's Preschool as their second campus in Laguna Beach, retaining some of the educational heritage from when it served as the Temple of Scientific Philosophy from 1934-1940. *Los Angeles Times* reports dating from 1971, 1987 and 1996, respectively, describe the school, still operating today, as being housed in a "17-room" "converted monastery" that "used to be a nursing home".

(photos: annelieseschools.com)

Photos on the website of what is now Anneliese's Preschool, where pupils are impressed with the values of love, freedom, self-discipline and eco-awareness, show many of the original features of the building mentioned in the Historic Resources Inventory (p.50), including the "casement windows" and painted tile walls and archways (above) and "painted tile stairway" (right).

LAGUNA HOME SOLD TO CHURCH GROUP

LAGUNA BEACH, Dec. 14.— Establishment of national headquarters of the Universal Progressive Christianity organization here has been announced following the sale of property formerly owned by M. Lalita Johnson. The property, located at 758 Manzanita drive, was formerly occupied by the Royal Order of Tibet, which is now a fraternity of Universal Progressive Christianity.

Daily classes in all branches of human life will start after the first of the year and service will be conducted every Wednesday, Friday and Sunday evening. The headquarters will be formally opened to the public December 20 at 8 p. m.

With establishment of national headquarters of the organization in Laguna Beach comes the announcement that the group's publishing department also will be located here and will publish the monthly magazine, "Universal Jewels of Life."

Santa Ana Register,
14 December 1935

Arrange Lecture At Laguna Beach

LAGUNA BEACH, May 29. — "The Invisible Man" is the title of a lecture to be delivered Friday night by Prof. George Adamski, of Pasadena, at the headquarters of the Royal Order of Tibet, located on Manzanita drive, according to a program announcement. He also will deliver a brief address on "Love and Jealousy." It also was announced that during the week addresses, supplemented by discussions, will be made by Paul Etzold.

Santa Ana Register, **29 May 1934**

Laguna Beach, Jan. 14 – Announcement of services by the Universal Progressive Christianity group at 758 Manzanita Drive in Laguna Beach was made today. Friday night at 7 o'clock the subject will be "Dr. Townsend's Third Party and His Pension Bill. What the Future Holds For It." Sunday night at 8 o'clock Ms. M. H. Weir will talk on "The Horoscope of Laguna Beach for 1888." Prof. George Adamski will speak on the Parliament of Religions which the group intends to bring to Laguna for the first session. Wednesday night Mrs. Alice Wells will speak on "Teachings of the Seven Masters." The "Universal Jewels of Life," a monthly manuscript, is given away free at all services.

Santa Ana Register, **15 January 1936**

Tibet Order To Conduct Festival

LAGUNA BEACH, Sept. 9. — A festival of inspiration will be held at the modern monastery of the Royal Order of Tibet September 10, 11 and 12, dramatizing the teachings of universal masters, to which the public is cordially invited. This is the first presentation to the public in the western world.

There will be a round table discussion with the purpose of bringing greater understanding of the present day needs in unifying all mankind under one supreme intelligence, it was announced.

If accommodations are desired reservations must be made. The monastery is located at 758 Manzanita drive, Laguna Beach.

Santa Ana Register,
9 September 1937

Wilderness Road, C., 1:45.
KPWB—G. Allison, 1; Philistine, 1:30; City Hall Broadcast, 1:45.
KFVD—Light Opera, R., 1:30.
KNX—Pete Pontrelli's Orch., 1 30.
KRKD—Judge Charles W. Fricke, 1:30; News, 1:45.
KGFJ—News, 1:45.
KFOX—Royal Order of Tibet, 1; Melody Sketches, 1:15; The Philistine, 1:30; Mrs. Lillian Culver, 1:45.

9 to 10 p.m.

KMTR—Floyd B. Johnson, 9; Montoya's Orch., 9:30.
KFI—Circus, 9; Inglewood Concert, 9:30.
KMPC—Have Party, 9; "Royal Order Tibet," 9:15; Mary's Melody Makers, 9:30; News, 9:45.
KEHE—Football, 9, 1 hr.

Los Angeles Times listings of broadcasting slots for the Royal Order of Tibet on KFOX (16 May 1936, top) and KMPC radio (20 September 1938, bottom).

In 1937 the Royal Order published *Petals of Life*, **a 16-page booklet with twelve poems "compiled by Professor G. Adamski".**

53

LONG-RANGE TELESCOPE ADDED TO LAGUNA PROJECT

READY FOR PUBLIC STUDIES

LAGUNA BEACH; April 29.— Forming an important part of study equipment at the Temple of Scientific Philosophy, located on Manzanita Drive, a six-inch Newtonian telescope from the Tinsley Laboratories at Berkeley has been installed.

Capable of long-range penetration and housed on a specially constructed platform in a garden corner overlooking the city and the ocean, the telescope, operated by Prof. George Adamski, is supposed to have the power of a twelve-inch instrument, with perfect clock-timing through which a star may be followed for hours.

The purpose of the telescope at this institution, Prof. Adamski explained, is to create an interest in the study of astronomy in conjunction with other scientific subjects. In this connection, the telescope will be open to the public on Sunday and Friday nights.

Prof. George Adamski is shown with Newtonian telescope installed at Temple of Scientific Philosophy at Laguna.
Times photo

Having installed his 6-inch telescope at the Temple of Scientific Philosophy in Laguna Beach "to create an interest in the study of astronomy in conjunction with other scientific subjects", Adamski clearly showed his interest in astronomy early on, perhaps already preparing for his mission to inform the world about the visitors from space. *(Los Angeles Times,* 30 April 1938)

The actual photograph was reproduced earlier in *George Adamski – A Herald for the Space Brothers* (2010).

Adamski robbed at gunpoint –
In its edition of Monday 10 July
1939 the *Santa Ana Register*
reports that two people were
arrested following the holdup of
George Adamski in Laguna Beach
the Saturday night before:

Both were booked at county jail
on robbery charges after they as-
sertedly robbed Prof. George
Adamski, lecturer and writer, as
he walked along the street in Laguna Beach about 8 p. m.

California Highway Officers Ben
Craig and Oscar Kelly were informed, according to reports, that
the pair drove up beside Adamski,
the Negro stopped the car and
Mrs. Jones jumped out and stuck a
gun against the victim, taking his
purse containing $16. Mrs. Jones
was booked at county jail at 9:35
p. m. Saturday and Grayson, at
8 a. m. yesterday.

Right: On 26 January 1940 the
Santa Ana Register reports that
George Adamski was leaving
Laguna Beach and the Temple of
Scientific Philosophy in March of
that year, to establish a spiritual
retreat centre in Valley Center, San
Diego County. From there he would
move to Palomar Gardens in 1944.

(Originally, a "dude –or guest– ranch"
was a ranch where visitors or tourists
could experience the nostalgia of
Western frontier days, which later
evolved into a kind of agritourism.)

PROF. ADAMSKI TO OPEN DUDE RANCH

LAGUNA BEACH, Jan. 26.—
Plans for the development of a
combination study and resort center with all the advantages of a
modern dude ranch for recrea-
tional activities, to be located on
an 80-acre tract of land in Valley Center, nine miles north of
Escondido, were announced today by Professor George Adamski,
lecturer and economist, associated
with the Temple of Scientific
Philosophy on Manzanita drive.
The project, including acquisition
of land and construction of the
first unit of building, represents
an initial outlay of about $80,000,
it was added.

Donating an Oriental touch to
the enterprise is the name selected for the place, Kashmir-la.
The tract is located off Colgrade
road, a mile from the town of
Valley Center and nine miles
from the newly planned Palomar
observatory.

Plans for the main building, to
be constructed of rock and con-
crete, call for a main dining
room, lobby, music and lecture
room, offices and 12 guest rooms.
Accommodations for guests also
will be provided for in cottages.
Other construction features include a power plant, swimming
pool, stables, garage and service
shops.

Forming an important part of
study equipment to be installed
at Kashmir-la is a 40-inch port-
able Cassegrainian telescope to
be constructed by the Tinsley
laboratories at Berkeley, Professor Adamski explained. At first
it was planned to have this instrument installed here in Laguna
Beach but under the new set-up
it will be shipped direct from
Berkeley to the Valley Center
project.

Professor Adamski is making
plans to leave here in the early
part of March for his new venture.

The Royal Order of Tibet – Chronology
Based on available data and newspaper reports

1932 – *The Invisible Ocean* published. First mention of The Royal Order of Tibet (p.23).

Classes held at Studio 102, 213 Broadway, Los Angeles, and likely elsewhere.

1933 – Lalita (Maud) Johnson establishes the Order of Loving Service and learns about Adamski and his teaching through *The Invisible Ocean*.

Adamski is invited to lecture for the Order of Loving Service at the Little White Church in Laguna Beach.

Royal Order of Tibet HQ moves to Hotel Castle Green, Pasadena, awaiting the availability of the property in Laguna Beach.

Classes taught at Hotel Castle Green as well.

1934 – January: Royal Order of Tibet opens new HQ at the Temple of Scientific Philosophy, 758 Manzanita Drive, Laguna Beach.

1935 – October: Mrs Johnson sells 758 Manzanita Dr. to Marguerite H. Weir who, in December, establishes Universal Progressive Christianity, of which the Royal Order of Tibet is now a fraternity with closed meetings.

1936 – Monthly *Universal Jewels of Life* handed out for free at services.

May: Royal Order starts 15-minute broadcasting slots on local California radio stations KFOX in Long Beach and KMPC in Los Angeles (until September 1938).

Wisdom of the Master of the Far East – Questions and Answers, Vol. 1 published. (This volume also announces a publication titled *The Seal of Mastery*, which doesn't seem to have materialized.)*

1937 – *Petals of Life – Poems* published.

Satan, Man of the Hour published. (Republished in *Flying Saucers Farewell*, 1961.)

1938 – April: 6-inch telescope installed at the site of the Temple of Scientific Philosophy, gifted by Mrs Maud Johnson.

1940 – January: Adamski announces plans for 'Kashmir-La' in Valley Center.

March: Adamski moves to Valley Center; Royal Order of Tibet effectively discontinued.

*Claims that *Wisdom of the Masters of the Far East* was later reworked and published as the *Science of Life* study course (1964) cannot be confirmed when reading these texts side by side. Perhaps it was the unpublished *The Seal of Mastery* that served as the basis for the *Science of Life* course.

PART II

THE SEA OF CONSCIOUSNESS

Gerard Aartsen

*"Consciousness is causal,
and physical reality is its manifestation."*

–Stephan Schwartz as quoted in
Ervin Laszlo, *What is Reality?* (2016)

THE SEA OF CONSCIOUSNESS – WHERE
SCIENCE VINDICATES ADAMSKI'S TEACHING

"At the young age of thirteen, endowed with a fervent desire to more fully understand these universal principles [the laws of nature and creation], George, along with a family friend and mentor, journeyed to Tibet. His training, under close supervision, consumed the next five years. The school hidden deep in the Himalayas, taught select students a rigorous curriculum ending in mastery of the four universal elements, Earth, Fire, Water, and Air. A test is associated with each phase to prove mastery before a student may proceed to the next step. For example, in order to complete the master of Air, the student must demonstrate the ability to levitate for extended periods of time. For Water, he must be able to control the molecules and form oxygen while breathing under water at the bottom of a lake. Similar tests apply to each of the other associated elements. Also, all students were very carefully instructed in the natural art of communication known as telepathy."[1]

Thus did Fred Steckling, the late president of the George Adamski Foundation, describe the training that George Adamski received in Tibet in his early youth in a profile he wrote for *The Contactees Manuscript* (1993).

While not unique, accounts of people who went on to become influential authors, teachers or healers after their training with the Masters of Tibet in hidden valleys or monasteries are quite rare; we only know about some of these through the work of H.P. Blavatsky[2], Murdo MacDonald-Bayne PhD[3], Rolf Alexander MD[4], and Baird T. Spalding, although his were out-of-the-body experiences.[5]

There is only one known instance where Adamski himself refers to his training in Tibet, when he is quoted in a *Los Angeles Times* report from 1934 about the Royal Order of Tibet: "I learned great truths up there on the roof of the world, or rather the trick of applying age-old knowledge to daily life, to cure the body and the mind and to win mastery over self and soul."[6] In *The Invisible Ocean* he seems to refer to his training when almost literally using the terms in the description above: "The fourth dimension is responsible for all form. It is a united condition of all elements, as we united earth, fire, air and water. The earth is visible, the fire is visible and the water is visible. The fourth element is the air, which is invisible." (p.12)

Comparing the invisible ocean of space to the physical oceans on Earth, Adamski writes that "between the floor of the ocean and the surface there are many strata of different vibration and that different creatures inhabit each strata" which implies that "there must be millions of different states of consciousness in which the ocean creatures are living" (p.4).

In the rapidly developing field of systems science Hungarian professor Ervin Laszlo, who was dubbed a "modern-day genius" in a 2017 PBS documentary, recently published his latest insights in two books, *The Intelligence of the Cosmos* (2017) and *What is Reality?* (2016). Systems science is an interdisciplinary field that studies the nature of systems, from simple to complex, and professor Laszlo states that "Consciousness is, and has always been, present in all clusters, from quanta to galaxies."[7] It is in the context of these cutting edge scientific insights that Adamski's *The Invisible Ocean* – and his later philosophical treatises – reveal their true significance.

Read with the understanding of 21st-century systems science, the quote from page 4 above clearly refers to the evolution of consciousness and the refining (spiritualization) of matter, as higher states of consciousness require forms of manifestation at higher rates of vibration for their expression. For students of the Wisdom teaching it goes without saying that this notion includes states of matter that, while physical, may lie beyond the range of the dense-physical matter that falls within our vision, just as there are frequencies of light and sound that escape our dense-physical sight and hearing.

In this regard, George Adamski himself has made seemingly

conflicting statements that are bound to be confusing if read without proper understanding. In one publication he first answers the question whether the Space People are "etheric spirits" by saying "No they are not. They are flesh and blood like you and me", while replying to a later question that "they can place their mind in a high frequency state that causes their body to become invisible to our limited range of vision".[8]

Adamski seemed to loathe the terms "ether" and "etherian", perhaps because he saw how in the 1950s these had opened the door to the spectre of people claiming to be "channels" for discarnate or excarnate beings in space. One could say there was a veritable invasion of such – some 'wishful thinkers', others opportunists – in the wake of the many cases of actual personal contact that all bore the promise of a better world if humanity could overcome the divisions it had erected and cooperate for the outward manifestation of its oneness.

Yet, at its frontiers, modern science is also beginning to see the reality of these subtler planes of etheric matter. In *What is Reality?* professor Laszlo, for instance, refers to historical precursors to the latest scientific notions: "A century before Plato, Pythagoras spoke of the 'aether' as the fifth element of the world, in addition to earth, air, fire, and water. Aristotle pointed out that this element does not have any properties; it does not become hot or cold, as the other elements do. (…) The rishis ('seers') of India named the deep dimension of reality 'Akasha'. A Sanskrit term, Akasha stands for the fifth element of the world, beyond *vata* (air), *agni* (fire), *ap* (water), and *prithivi* (earth). Akasha holds all the elements within itself, but it is also outside of them. (…) The Akasha is an intuitive, yet remarkably up-to-date concept of a deep dimension underlying the universe. At the dawn of modern science this dimension was considered to be a space-filling substance, the ether."[9]

We may be forgiven for thinking it is exactly this 'deep dimension' that George Adamski described in *The Invisible Ocean*: "There is no division anywhere. Fire is in the earth; water is in the earth; all interblending, the atmospheric air is right on the surface, interpenetrating into the inner layers, and also interblending. When you realize these things, you begin to see the invisible or real kingdom of the universe, and then you call it the fourth dimension, but it is not new, and it is not a dimension, for it takes them all to bring forth a manifestation." (p.13) "The manifestation is the fifth." (Top of p.13)

The term "etherian" to denote the subtle-physical nature of the space visitors and their craft was first given prominence by Meade Layne, then-president of the Borderland Sciences Research Association in California and himself a long-time student of the UFO phenomenon. In his article '"Mat and Demat" – Etheric aspects of the U.F.O.' (where "mat" and "demat" stand for "materialization" and "dematerialization") Mr Layne came very close to explaining the etheric planes of matter for what they are, with a definite hint of the underlying theosophical notion about matter when he says that "the prototypes of all our metals exist in etheric matter".[10]

Indeed, as the Tibetan Master Djwhal Khul stated via Alice A. Bailey, the etheric "is the true form to which all physical bodies in every kingdom of nature conform".[11] Even in 1936 Adamski wrote: "The visible forms come forth from the invisible essence of God, the indestructible substance of the universe. Consciousness molds all form out of this primal substance, first in the invisible state and then gradually bringing it into a state of greater density through a descension of vibration in the composing elements until it is finally brought into the extreme coarse state of vibration known as visible manifestation."[12]

Hence, the visibility of the bodies and craft of the visitors is not so much a matter of materialization or dematerialization, but of the rate of vibration of the subatomic particles of which they are composed – still matter, but of a subtler nature. As Adamski was told by his contacts, aeroplanes "could fly blindly into our ship without seeing it. If we permitted you to come as close as that, when you hit, you would find our craft as solid as though functioning in a lower frequency."[13]

Another contactee featured in *The Contactees Manuscript* quoted from at the opening of this chapter, is the late Scottish esotericist

Benjamin Creme. Speaking about his visit to George Adamski's lecture in London in 1959, he says: "He [Adamski] confirmed that the space people and their UFOs are not solid physical matter. Their solid physical form is a temporary manifestation. They are really of etheric physical matter, above the level of our physical mass."[14]

Earlier, in his first publication that signalled his interest in or, more likely, his knowledge of extraterrestrial life, *The Possibility of Life on Other Planets* (1946), Adamski himself said: "Even upon planets whose atmosphere is so rare that life seems impossible there may be intelligent forms existing – forms having the power of reason such as we possess, but the actual physical construction may be so fine as to be almost invisible to our sight, limited as it is to this particular plane of manifestation."[15]

Three years later he published *Pioneers of Space*. While it is true that this volume was presented by the author in his Foreword as being 'in the field of fiction'[16], he sets the stage for his space travelogue solidly on the basis of the scientific facts and data of his day and, perhaps more meaningfully to the discerning reader, dedicates his book "to the Goodwill of Man", in clear congruity with the message from his extraterrestrial contacts in later publications.

In the same vein, the striking similarities between descriptions in *Inside the Space Ships* (1955) and *Pioneers of Space*, which detractors use to dismiss all of his work, may also be an indication that the latter book was meant to test the public reaction to the idea of life on other planets, which he had experienced but was not allowed to write or speak about at the time. Even without Benjamin Creme's confirmation of this notion through his own contact with a Master of Wisdom[17], this would not be a stretch if we recognize to what lengths the visitors themselves go in respecting our free will as they try to make their presence known. Also, all contactees are asked to leave their audiences the freedom of thought to decide for themselves if they want to take seriously the reality of the extraterrestrial visitors.[18] It is in this spirit that Adamski said he would leave "each man the privilege of believing or disbelieving, of benefiting from a higher knowledge, or casting it aside in derision and skepticism".[19]

Interestingly, Adamski makes a clear reference to *The Invisible Ocean* in his Foreword to *Pioneers of Space* when he writes: "Is the ocean near the surface void of fish because it is light pressured? No. Is it void of fish because of a heavy pressure at the bottom of the ocean? No." And in what could be seen as an allusion to the universality of the human form throughout cosmos: "And all between, does the pattern of the fish change due to the heavy pressure at the bottom or the light pressure at the surface of the ocean? No. The pattern is the same. The only difference is in the structure of the form itself, complying with the various pressures."[20] This passage uses the exact same imagery as the text on page 4 in the current volume, as does the opening of Lesson Five in his *Science of Life* study course: "We are actually living in an invisible sea of life. And as I have said before, we should school ourselves to become aware of the visible and invisible at the same time. For today we are only aware of the visible forms with which we have contact. Yet all forms rise from the lowest upwards, looking so to speak into invisible space. And why should this be so? Does not all creation look to its creator as a child looks to the mother for guidance? And as space is the birthplace of all forms, they look to and live within the household of their birth."[21]

When we suggest that George Adamski's training in Tibet to control the four elements and "win mastery over self and soul" likely included being trained to leave the dense-physical body at will for space travels in the etheric body, we are not simply speculating. (Another fascinating example can be found in Vera Stanley Alder's autobiography *From the Mundane to the Magnificent*.) As a matter of fact, Adamski wrote about conscious out-of-body travel in another lesson of his *Science of Life* course: "All planets and form bodies are born from the elements of space invisible to sense sight but not to consciousness". Here he also writes about "conscious travelling", i.e. travelling outside the body, which could also take the form of remote viewing or, in an advanced stage, bilocation. In other words, he says, "a sort of telecasting connection is made between the consciousness and the mind and time and distance do not mean anything for the mind is interested in the scenes that the consciousness is bringing to it." "We must remember that consciousness is the sea of life within which all forms are living regardless of what they may be." Referring to his lectures for the Royal Order of Tibet, he writes: "Many times when I was teaching in the early thirties a student would be sick and would not be at the class, yet on the following week when they returned he would report that he had not missed the class. For I was at the bedside giving the instructions through my mind and body while at the same time my consciousness was at the bedside of the one who was sick. (...) In one place I was a solid form and in the other a thought form."[22]

In a letter dated 16 August 1950 Adamski wrote about his experiences in *Pioneers of Space* to a Mrs Emma Martinelli: "In this letter I have explained, using illustrations, how one may venture from one place to another, while his physical is in one place and he is in another. That is the way I have written this book. I actually have gone to the places I speak of; I actually have talked to the ones I speak of. To you I can reveal this since your letter reveals much [understanding], while to others I keep silent about this." Later, in a letter of 16 January 1952 (before his meeting with Orthon in the desert in November that year) he explained: "... speaking of visitors from other planets, you see, in the physical I have not contacted any of them, but since you have read Pioneers of Space you can see how I get my information about these people and their homelands."[23]

In a later article he elaborated: "Most human bodies of Earth could not endure extended journeys for there are very few who blend their consciousness with All-Consciousness (...) for this involves a sacrifice of one's personality or ego (sense-mind) unto the dictates of consciousness where the mind is servant unto its Creator..."[24]

Here we see how Adamski explains that his experiences aboard space ships and on other planets still require a vehicle to express and experience Life – and the higher the form of manifestation, the higher the vibration of its vehicle.

Charting a "new map of consciousness" in his book *What is Reality?* professor Laszlo provides scientific evidence, gathered from many sources, of out-of-body experiences, remote viewing, and other instances of consciousness retaining its individuality and sentient activity outside the dense-physical body, bringing him to conclude: "Consciousness beyond the brain was previously an esoteric assertion, but it now has a scientific foundation."[25] This takes Adamski's claims straight out of the realm of fantasy or wishful thinking and brings even his much ridiculed account of his three-day trip to Saturn in 1962 into the stratum of the possible. (See also p.89ff about other planets being inhabited.)

In his efforts to prove his protegé's claims the Adamski Foundation's current president, Glenn Steckling, insists that all of Adamski's space travels could only have been strictly physical, suggesting he knows better than the latest findings in neuroscience that support how consciousness is seen in the Wisdom tradition, which Adamski was trained in. However, science itself is now acknowledging that "spacetime", the dense-physical dimension of reality, could not be the entire reality of the universe: "Until recently, physicists considered the complex plane [the dimension beyond space and time] a mere mathematical device, useful for predicting the behavior of quantum systems over time, but not an element of physical reality. This view is being reassessed. The complex plane appears to be a dimension or domain of the physical world beyond spacetime."[26]

Desmond Leslie addresses the debate about the physical reality of Adamski's experiences in his commentary on George Adamski for the revised and enlarged edition of *Flying Saucers Have Landed*, where he comments that some of these had "the marks of a spiritual, out of the body experience". As a student of the Wisdom teachings he understands that this would make such accounts "no less true because of this. It is the Consciousness, not the body that is the Self."[27] Scientific research into the nature of consciousness is beginning to catch up, with neuroscientist Stephan Schwartz going so far as to say that "Consciousness is causal, and physical reality is its manifestation."[28] So maybe, Leslie writes, Adamski's "Mandate was to try and establish only the objective reality of the visitors; a difficult enough task in itself, without confusing the layman with anything remotely esoteric". Nonetheless, Adamski did admit to Leslie "that no one of us could be taken to another planet in our system and see the home world of its inhabitants in *his present bodily form or condition*."[29]

In recent years several publications have provided further proof and evidence of the authenticity of George Adamski's experiences, photographs, and claims of contact with extraterrestrial visitors.[30] Earlier, Desmond Leslie had already summarized several examples of the accuracy of Adamski's descriptions in *Inside the Space Ships*:

(1) The bands of radiation surrounding the earth, which were later discovered by science and named the 'Van Allen Belts';

(2) The 'fireflies' in space, that were later sighted by astronaut John Glenn;

(3) A patch of light, into which Adamski saw a space craft enter, was later also described by astronauts on the Gemini 6 and Mercury 9 missions.[31]

While it is certainly helpful to have the facts available that counter the many false claims and outright lies meant to discredit Adamski, these recent efforts still fall short of heeding his admonition: "Too much stress has been laid on the phenomena of sightings, and not enough on the pathway of life these visitors have shown us. It is my opinion, that *were all groups to couple their research into sightings with a study of the information given by the space people [...] an expanding mode of living would open before them*."[32] [Emphasis added.]

The "information given by the space people" in the original 1950s sources at its core is concerned with consciousness – more specifically, the need to expand our consciousness to the extent that man, in Adamski's words, "must look upon his fellow being as himself, the one a reflection of the other".[33] This, we are told, is the only way to overcome the dangers to the human race and the planet itself that result from the divisions inherent in our current way of life. In terms of the latest insights from systems science: "The evident purpose of evolution in this [spacetime] domain is to achieve coherence in the domain of natural systems, and embracing oneness in the sphere of consciousness."[34]

In practical terms, this simply means that as individuals we need to realize that those who fear for their sense of security because migration is changing the world they know, or who have never known a sense of security because of oppressive economic circumstances in their own land; those who have left their family, friends and community in pursuit of liberty; those waiting in the Spanish enclave in Morocco, at the US border in Mexico, in the detention centre on the island of Nauru near Australia, or the miserable refugee camps in the Middle East and Greece for a chance to live a life without hardship or conflict; that all of those people are us – *they are us.* And if we don't see that the basic needs of every man, woman and child are identical, the onus is on us to expand our sense of humanity to include 'the other', to venture into the stratum of the ocean of consciousness directly above the one where we currently feel at home.

When this information was first given, during the Cold War, the ideological battle that caused the divisions among humanity expressed itself in the political field between the Western capitalist ideal of "freedom" and the Eastern communist ideal of "justice", as if these are mutually exclusive concepts. So if governments and the military were to retain public support for their particular ideology and the arms race necessary for its "protection", it is no mystery why they felt impelled to contain and reverse the growing public interest in messages about international cooperation that appealed to the brotherhood of man, by gradually infecting the field of Ufology with disinformation.

With the collapse of the failed communist experiment, from the 1980s onward the divisions began to manifest in the socio-economic field. While the people were promised greater freedom through 'free markets', it was the economic and financial powers who were handed near-unlimited freedom, which triggered a steady and relentless attack on the social, civil, and human rights. These have been secured through the rise of the social movements that sprang up since the 19th century as a result of humanity's gradually growing awareness of itself as One, but now stand in the way of maximizing profits.

While claims of ET contacts were dismissed because of the ridicule bestowed on the accounts and character of the original contactees by the establishment of their day, likewise religions were dismissed by thinking men and women because of the separative dogma that obscures the original teachings. A comparative study of the essence of the world's major religions, however, shows that at the core they all teach that at the beginning or end of each cosmic cycle a Teacher returns with a new revelation about the source and evolution of consciousness, which needs to be given expression by living according to the Golden Rule to establish right human relations.[35] In the analogy offered in *The Invisible Ocean*, if we are "gradually drawn up in [our] evolution" (p.4) we must, each of us, "destroy the dividing line" (p.7) in our consciousness, between our lower nature and our higher Self, which in turn will remove the dividing line that separates us from others and their needs if, Adamski said in 1932, we learn "to understand [our] fellow man, and how to serve them" (p.7). Likewise, the latest scientific findings have led professor Laszlo to conclude: "The evolution and persistence of systems in nature is rooted in cooperation rather than in competition."[36]

George Adamski seems to acknowledge that this is such a time in history when he says, "the time is at hand for the co-operation of souls seeking the highest, and to be bound together only by the ties of the Infinite Love and Wisdom" (p.23). And indeed, when we broaden our view we see that, since 1875, the world is being re-acquainted with the spiritual facts of life with the re-introduction of the Ageless Wisdom teachings through H.P. Blavatsky; the announcement, through the work of Alice A. Bailey, of the externalization of the spiritual kingdom in nature, the hierarchy of Masters of Wisdom; and the pronouncement, through Benjamin Creme's mission, that the World

Teacher for the New Age, the Master of all the Masters, has physically re-entered the modern world in 1977 and awaits the time humanity is at its most receptive for his Declaration to inaugurate the cosmic cycle of Aquarius, with its keynotes of unity and synthesis. According to Creme, it is to ensure the smoothest possible transition into this new dispensation that the Elder Brothers of humanity are being assisted by an unprecedented number of Space Brothers at this time in our history. And even if we disregard these esoteric postulations, the information coming from the space visitors chimes in perfect harmony with the essential teachings given to humanity at the beginning of every cosmic cycle, regardless of cultural or historical setting.

Despite the title of what we now know was his second booklet, *Wisdom of the Masters of the Far East*, and his various references to the Men of Wisdom in his later writings, some Adamski adepts dispute that he trained with the Masters of Wisdom in Tibet, insisting that it was really the Space Brothers. They seem to think of 'Masters of Wisdom' as an undesirable mystic notion, rather than the logical, natural next 'stratum' in the manifestation of a planetary Life, as consciousness evolves and ascends through the successive kingdoms in nature. Of course, one needs to accept that the human kingdom is not the culmination of the process of evolution in order to see this.

Since Adamski himself referred to some highly evolved beings from other planets as 'Masters' in *Inside the Space Ships*, it wouldn't even matter if the Teachers he trained with in Tibet were Masters from Earth or from another planet – they had clearly evolved beyond the strictly human kingdom to be able to teach him (as well as others) what they did. So whether from Venus, Saturn, Earth or somewhere else, we can safely state that Adamski trained with the Masters in Tibet.

It should come as no surprise, then, that even his earliest writings concern the evolution or expansion of consciousness, which is the essence of all true Wisdom teachings. Therefore, rather than "a convoluted circus of misinformation and misdirection", as Mr Steckling calls any connection that is made between Adamski's information and the Ageless Wisdom teaching[37], we would have to deliberately close our eyes to what Adamski himself wrote and taught throughout his life, so as not to see consciousness as the cause of the material manifestations that we experience, or the innate drive, indeed the need, to expand our consciousness. Then it would also be of no avail to repeat what has been pointed out above, that Adamski's own allusions to a physical reality above that which falls within our visual spectrum are increasingly finding support in the work of the world's most advanced scientists (p.89ff). As professor Laszlo says: "[T]he physical world is a domain, a segment, and hence a manifestation, of the intelligence of the cosmos."[38]

More than any other book about contact with people from space visiting Earth, George Adamski's *Inside the Space Ships* shows that the visitors who are here to help humanity with our transition into the new cosmic cycle at this time in our planetary history hold before us the exact same message: Life is about the expansion of consciousness, and the only safe path forward for humanity is to manifest our growing consciousness by living by the higher ethic of the Golden Rule – treating our fellow human beings as we ourselves want to be treated. In *The Intelligence of the Cosmos* Ervin Laszlo leaves no doubt that this message can also be found in the new paradigm that emerges from the forefront of science: "The hallmark of the evolved forms of consciousness is the mind-set that emerges in ethical, insightful, and spiritual human beings."[39]

Whether we feel attracted to any particular way this message is expressed or presented would matter not the slightest if indeed we would take it to heart, rather than idolize the messenger or any particular Teacher. Let us accept, then, that we are one human family who share the same planet and that, far from unique, we are a very common, not to say universal manifestation of Life that expresses itself in evolving and successive grades of consciousness. As a teacher in the Wisdom tradition, in *The Invisible Ocean* Adamski already expounds exactly this notion of "the many different states of consciousness in the one body called the universe or GOD" (Preface, p.1). The only difference, he says, is in vibrations: "Finer vibrations are brighter and better because the pressure is less heavy. At the bottom, things are coarse and heavy because the pressure is strong." (p.7) But, "through study and understanding we learn to bring the superconsciousness into harmonious contact with the conscious mind, thereby clearing the way for the Expression of Divine Mind" (p.13). Remarkably, we now find this confirmed in 21st-century science: "Through the action of love, an evolved consciousness fuses the elements of the mind into a higher unity. Attaining this 'spiritual evolution' is the meaning of existence."[40]

Says professor Laszlo: "The body and the brain, and the organism as a whole, receive and resonate with the intelligence that permeates the universe."[41] Or, in the words of neurosurgeon Eben Alexander, quoted by professor Laszlo: "There is a profound Intelligence or Source underlying creation and evolution of the universe from which all things originate and to which all things return."[42]

In *The Invisible Ocean* George Adamski put it this way: "As we grow in consciousness, we get nearer and nearer the light and finally we see the full light, just as the fish can begin to see the light when it is within fifteen feet or so from the surface." And, "The fish that finally advances to the highest strata finds himself on the surface and sees the light of the world for the first time." (p.5) Note that the phrase used here is not "the light of day" but "the light of the world". This, of course, is a clear reference to the Christ as the symbol or personification of the Christ consciousness, the divine spark in every human being. And it is this 'Christ consciousness' or higher Self that, according to the Wisdom teaching, seeks ever more perfected expression through its personality vehicles in incarnation, in its service to the Plan of Evolution to spiritualize matter.

While mainstream science and its more stubborn adherents insist that empirical evidence alone can help us understand life, let us acknowledge that understanding life is about our *own* understanding of life, which changes with the expansion of *our* consciousness, and is therefore about the evolution of consciousness *itself*. So whether we are studying the constitution of the Cosmos or the information coming from the Space Brothers, the essence of the world's religions or the Ancient Wisdom teachings, we are essentially talking about the evolution of consciousness as the manifestation of the One Life that underlies everything in creation.

With astrophysics moving closer to the broader understanding of reality that quantum physics and systems science offer, it should not come as a shock to know that Ervin Laszlo himself admitted he has had access to information about extraterrestrial visitors, based on which he asserts: "I am certain that contact has been made."[43]

Presenting the universe, as professor Laszlo does, as a "sea of coherent vibrations"[44], science has now arrived at the point of understanding where Adamski's teaching started in 1932 – "the invisible ocean of vibrations" (pp.10-11). And as the findings of avant-garde scientists blaze the trail for their mainstream counterparts, the significance of Adamski's work will gain wider acknowledgement, and no doubt recognition of the authenticity of his contacts and experiences with it.

References

1 Fred Steckling, 'Chapter 1: George Adamski'. In: Joseph Randazzo (ed., 1993), *The Contactees Manuscript*, p.1

2 Sylvia Cranston (1993), *HPB – The Extraordinary Life & Influence of Helena Blavatsky, Founder of the Modern Theosophical Movement*, p.80ff

3 Murdo MacDonald-Bayne Ph.D. (1938), *The Higher Power You Can Use*, pp.9-10

4 Rolf Alexander M.D. (1946), *The Voice of Talking Valley*, Chapter VIII, 'The Trappa'.

5 Baird T. Spalding (1924), *Life and Teaching of the Masters of the Far East* and Benjamin Creme (1987), *Share International* magazine, Vol.6, No.4, May, p.18

6 'Lamaistic Order to Be Established Here', *Los Angeles Times*, 8 April 1934 (see pp.46-47 in the current volume)

7 Ervin Laszlo Ph.D. (2017), *The Intelligence of the Cosmos*, p.22

8 George Adamski (1965), *Answers to Questions Most Frequently Asked About Our Space Visitors and Life on Other Planets*, p.10 and p.16

9 Laszlo (2016), *What is Reality?*, pp.16-18

10 Meade Layne (1955), ' "Mat and Demat" – Etheric aspects of the U.F.O.', *Flying Saucer Review*, Vol.1, No.4, Sep-Oct, pp.22-23

11 Alice A. Bailey (1950), *Telepathy and the Etheric Vehicle*, p.139

12 Adamski (1936), *Wisdom of the Masters of the Far East*, p.24

13 Adamski (1955), *Inside the Space Ships*, p.156

14 Edward T. Foster, 'Chapter 10: Benjamin Creme'. In: Randazzo (ed.; 1993), op cit, p.104 (see note 1)

15 Adamski (1946), *The Possibility of Life on Other Planets*, p.20

16 Adamski (1949), *Pioneers of Space*, p.6

17 Benjamin Creme (2007), 'Questions and Answers', *Share International* magazine Vol.26, No.3, April, p.27

18 See e.g. Gerard Aartsen (2015), *Priorities for a Planet in Transition*, pp.6-8 and throughout.

19 Adamski (1955), op cit, p.127

20 Adamski (1949), op cit, pp.4-5

21 Adamski (1964), *Science of Life* study course, Lesson Five: 'Consciousness, The Intelligence and Power of All Life'

22 Adamski (1964), op. cit., Lesson Eleven: 'Exploration of Cosmic Space'

23 Håkan Blomqvist's blog (2013), 'The George Adamski correspondence'. Available at <ufoarchives.blogspot.com/2013/10/the-george-adamski-correspondence.html> [Accessed 11 May 2018]

24 Adamski (1962), 'How to Know a Spaceman if You See One'. In: Gray Barker (1966), *Book of Adamski*, p.53

25 Laszlo (2017), op cit, p.24

26 Laszlo (2016), op cit, p.20

27 Desmond Leslie and George Adamski (1970), *Flying Saucers Have Landed*, Revised and Enlarged Edition, pp.258-59

28 As paraphrased in Laszlo (2016), op cit, p.40

29 Leslie and Adamski (1970), op cit, p.259

30 See Michel Zirger (2018), *Authenticating the George Adamski Case*; Rene Erik Olson (2017), *The George Adamski Story*; and Glenn Steckling (2016), 'Prologue 2016' in George Adamski (2016), *The Extraordinary People, Power and Events Behind the Flying Saucer Mystery II*.

31 Leslie and Adamski (1970), op cit, p.242

32 Adamski (1957-58), *Cosmic Science for the Promotion of Cosmic Principles and Truth*, Series No. 1, Part No. 3, Question #58

33 Adamski (1955), op cit, p.239

34 Laszlo (2017), op cit, p.46

35 Aartsen (2018), 'A global crisis in consciousness ... and the age-old Laws to guide us'. *The Edge* magazine, 1 October. Available at <www.edgemagazine.net/2018/10/a-global-crisis-in-consciousness/> [Accessed 28 February 2019]

36 Laszlo (2017), op cit, p.38

37 Glenn Steckling, 'Prologue 2016' in George Adamski (2016), *The Extraordinary People, Power and Events Behind the Flying Saucer Mystery II*, p.54

38 Laszlo (2017), op cit, p.22

39 Ibidem, p.39

40 Ibid., p.41

41 Lazlo (2016), op cit, p.45

42 Ibid., p.39

43 Interview with Ervin Laszlo in Tessa Koop (2008), 'Tessa Koop meets...'. The original interview is no longer online, but the relevant fragment from the interview can still be seen in the video compilation 'End the UFO/ET Disclosure nonsense!' (2015), available at <www.youtube.com/watch?v=U78n64c8K7A> at the 3-minute mark.

44 Laszlo (2017), op cit, p.21

SUFFER THE LITTLE FISH
(AT THE BOTTOM OF THE OCEAN)

Diligent research of voter lists and US census data has established that from 1919, after he left the US Army, George Adamski worked as a manual laborer in various occupations in different parts of the country, before settling in the Los Angeles area around 1928 to teach philosophy.[1]

In his early days as a teacher Adamski travelled about California, New Mexico, and Arizona as a self-styled "wandering teacher", giving lectures and teaching seekers of esoteric knowledge about the Universal Law which he himself had studied for five years with the Masters of Wisdom in Tibet, from around 1904 until 1909.

He soon founded The Royal Order of Tibet whose meetings were convened in Los Angeles and nearby Pasadena until at least 1933, while he also taught classes for the Order of Loving Service that had been established in Laguna Beach by Mrs Maud ('Lalita') Johnson. She provided the funds for the acquisition of the Laguna Beach property at 758 Manzanita Drive where the Temple of Scientific Philosophy would be established in late 1933 as the monastery for the Royal Order of Tibet. Here, from January 1934 onward, he conducted lectures and informal discussions, which were also published in tracts and booklets.

Three years after it was published *The Invisible Ocean* received a scornful review in a Theosophical publication (see next page) which, in hindsight, might be seen as an omen of the accusations and public derision that awaited George Adamski later in life and that continue to taint his reputation to this day. No doubt, detractors will think they find their suspicions confirmed in this malicious review, but when we take a closer look, we find that the writer does not identify

THE O. E. LIBRARY CRITIC

Published monthly at 1207 Q St., N. W., Washington, D. C.

BY

The O. E. Library League

| Vol. XXIII | March—April, 1935 | No. 6 |

Yearly subscription. United States and foreign, fifty cents. Single copies, five cents. British and Canadian postage stamps, paper currency and blank (unfilled) postal orders accepted.

The Sacred Oil Man.—One George Adamski has attracted quite a following in Los Angeles by his lectures on occult subjects. Besides lecturing on occultism, so-called, he has been peddling an oil consecrated by the Masters of Tibet, guaranteed to remove facial blemishes, restore vitality and even, so I hear, to raise the dead; two dollars an ounce bottle, if you please, as long as the supply lasts. It is said that the ladies just fell for it. Adamski represents "the Royal Order of Tibet", which "is for the purpose of establishing the All into One Eternal Life Progress", so he says. That sounds sweet and we need not be surprised that George is, or was, collecting funds for building a two million dollar monastery at Laguna Beach. The funds were secured in exchange for stock in purported oil wells in Oklahoma. I learned much through reading Adamski's pamphlet on "The Invisible Ocean", among other things that "Hydrogen and Oxygen is boundless and limitless." I learned still more, namely, that the more damned nonsense one talks the surer one is of finding followers, especially when backed by bottles of consecrated oil and a vigorous application of the Tibet racket.

I am told that Adamski's followers consist largely of disgruntled Besantite theosophists. Why, I know not, unless it is because, while Leadbeater's consecrated oil, or chrism, smeared on the top of the head, only purifies the soul, Adamski's oil removes wrinkles, which is much more important. The soul, being eternal, can wait to be purified, but wrinkles, no—they must be taken in hand at once.

Doubtless there is a germ of truth in the Adamskiite teachings. This attracts people who regard it as proof of the veracity of the hokum. It should be perfectly obvious that no occult adventurer could succeed if he put forth nothing but nonsense. The germ of truth serves as the bait, especially if ornamented with claims about Tibetan authority, and ultimately the sucker, or at least his money, is landed in the adventurer's pocket. Don't think that presenting a truth is sure proof of honesty or sincerity.

From: *The O.E. Library Critic*, Vol.XXIII, No.6, March-April 1935, edited by Henry Newlin Stokes (1859-1942).

'O.E.' was the Oriental Esoteric Center, based in Washington DC. Mr Stokes was an adherent of the 'Back to Blavatsky' movement, whose members felt the Theosophical Society under Annie Besant had strayed too far from the original teachings of H.P. Blavatsky.

any sources to back up his claims about Adamski's actions or motives. Reading this brief review we notice how the critic relies on hear-say constructions as the basis for his accusations: "so I hear", "it is said", "I am told". We may safely assume that nowhere else in the official files and archives that have been doggedly scrutinized for 'dirt' to prove that Adamski was a conman or a hoaxer have hints of him selling snake oil been uncovered, or these would already have been gratefully used in the many attempts to discredit him.

Yet, these allegations, unfounded though they are, do remind us of the few instances that hint at Adamski's healing abilities. In *The Pawn of his Creator* Henry Dohan writes: "The people who were around Adamski and who knew of his unusual powers were asked to keep it a secret. Adamski thought it would prejudice his prestige as a teacher, as people would take him for a magician."[2] With *The Invisible Ocean* we now have one instance where Adamski himself spoke of such powers, when he gives an example of overcoming conditioning: "I cured a lady of a paralytic condition" by cancelling out the conditioning notion of her being paralysed (p.20).

While his statement is of course no proof in itself, we do find a description in Maud Johnson's undated book *The Sacred Symbol* that provides an interesting correspondence. It seems Mrs Johnson suffered from obesity, "the after-effects of the unwise removal of a gland", and "very drastic dieting and starving and some exercising" were of no avail to get rid of her "superfluous flesh". She continues: "I was discussing various matters with a spiritual student and teacher one day and simply, in the course of conversation, stated that I had become thoroughly convinced that I could never get rid of my flesh through diet. It would never disappear unless I grew a new gland in place of the one removed, and I did not know how to do that. He replied: 'That is easy. Never doubt that you are growing one.' My whole being accepted it, then and there, and I said 'Oh, of course.' I never thought of it again. The idea simply became a part of myself. And from that moment on my body began to change. The change was gradual but continuous. In about two years I dropped from 172 to 128."[3]

While she doesnt't identify Adamski as the one who dropped the unconditioning thought in her mind, the correspondence between her description and Adamski's texts (see again p.20 and also p.28) makes it most likely. Also, independent confirmation of Adamski's extraordinary powers comes from Lou Zinnstag who, even long after distancing herself from Adamski, described how he diagnosed the cause of the deformity of a boy's eye while in Rome in 1963.[4]

To be sure, claims of such 'magic' healing powers will be the ultimate proof of his deceit for those whose view of reality is limited to what they can see and touch for themselves. Taken in the broader, spiritual sense, however, the word 'magic' means the 'Great Life' or divine life[5] – life as Adamski understood it, as shown in the previous chapter.

His critics refuse to let the facts get in the way of their accusations that Adamski would do anything for money or fame, yet his earliest writings show how consistent he was in his interest in consciousness and space, and the context provided in this volume shows how far ahead of his time his teaching really was. So, just as the unfounded claims of Adamski selling snake oil would expose an irreconcilable chasm between his character and motives on the one hand, and his teachings and philosophy on the other, this latter explanation, with eyewitness accounts backing up his own words, opens up a vista of consistency that cannot be denied.

However, the pattern of repeating accusations without factual basis has been a recurring feature in the attempts to denounce his relevance. With the commercialization of the world wide web and its attendant click-baiting, no amount of corrections or fact-checking seems adequate to stem the harebrained parroting of lies, half-truths and libel that have been hurled at Adamski, in many cases for the very reason that such detractors like to accuse the subject of their scorn of: making a name, fame or money. The only reason it was decided to include the *O.E. Library Critic*'s review of *The Invisible Ocean* here in full is to preclude the appearance of our being selective as an excuse to dismiss our argument. With the full text of *The Invisible Ocean* made available here, placed in the context of the insights of cutting edge systems science, it becomes apparent that a lack of actual facts, limited understanding and a generous helping of ill-will suffice to concoct long-lasting slander.

Admittedly, Adamski's character didn't always help his case. One well-known example that continues to rear its head is the claim that he gave away his true motives in a spell of alcohol-induced candour to a few young UFO enthusiasts visiting him at Palomar Gardens in 1956 or 1958 – several versions of this account exist, with the oldest in print dating back to 1978: "...during the Prohibition I had the Order of Tibet. It was a front. Listen, I was able to make the wine. You know, we're supposed to have the religious ceremonies; we make the wine for them, and the authorities can't interfere with our religion. Hell, I made enough wine for half of Southern California. In fact, boys, I was the biggest bootlegger around ... If it hadn't been for that man Roosevelt, I wouldn't have [had] to get into all this saucer crap."[6]

We cannot discount the possibility that George Adamski did indeed make those or similar remarks. Several accounts from even some of his close associates, like Desmond Leslie, show how Adamski was sometimes known to express himself in ways that lacked the subtlety of his teaching. Some of the adjectives that Leslie used in just two consecutive sentences to describe Adamski are revealing in this respect – "enigmatic fascinating, at times infuriating" and "lovable, provocative, evasive at times; and at other times overshadowed by a profundity that was quite awesome". Leslie continues: "You had to get him alone and relaxed to discover this deep inner Adamski....one often had the impression there were two people in that fine leonine body, the little Adamski, the burbler which always shoved its way to the foreground when the crowds gathered, talking non-stop...Then there was the big Adamski, the man we came to know and love, who appeared only to his intimates, and once having appeared, left them in no doubt they had known a great soul. The Big Adamski spoke softly with a deep beautiful voice, incredibly old, wise and patient. Looking into those huge burning black eyes one realised that this Adamski had experienced far more than he was able or willing to relate."[7] And in his obituary for Adamski, Leslie wrote: "I several times glimpsed the presence of a Master, and I was always sorry when the curtain came down again and the worldly mask obscured him."[8]

Readers who want a better understanding of Adamski's character, and the pressure he operated under, have a rich source in Tony Brunt's *George Adamski – The Toughest Job in the World*. It should be noted here that Mr Brunt should probably not be regarded a devoted 'believer', but he distinguishes himself by his sincere willingness and effort to understand the complexity of his subject's character. He writes: "Adamski also used knock-about humour as a leveler in masculine company, the macho combination of exaggeration and self-deprecation. In 1958 he told two visitors to Palomar Terraces that the Royal Order of Tibet (...) had been a racket to get around Prohibition..."[9]

And so it is equally possible that a UFO fan in his late teens visiting a famous 67-year-old contactee, teacher, writer, speaker, and philosopher in 1958 had no idea of the psychological or social pressure under which Adamski was living and working. Nor why he would need such boisterous bouts of venting the relentless stress that resulted from the public interest, the media scrutiny, and the scientific derision that he had to endure for his claims of contact with visitors from space. And then we don't even know what additional pressure was brought on by aspects of his mission that he was not allowed to speak about at all. To quote Desmond Leslie again: "As one of his intimates later told me, 'If George had been allowed to tell all he knew, his life would have been much easier for him. He'd have been able to prove his case.'"[10] Here Leslie continues by citing several instances where Adamski had revealed information in his accounts that were only confirmed by science later, concluding that "Adamski had too many 'lucky guesses' for comfort."[11] (See also p.67.)

In his book *Alien Base* (1998) UK researcher Timothy Good concluded that, "Apart from my own prejudices, I feel it is important to re-emphasize that a great deal of what Adamski spoke and wrote about the 'space people' and their technologies is now, on the verge of the twenty-first century, more plausible and more scientifically relevant than it was some 40 years ago."[12]

My own study of his teaching showed that Adamski was hardly the 'crackpot from California' that many made him out to be, but rather "a visionary teacher who was far ahead of the small minds that were waging a cold war against not just political opponents, but against the *fact* of our interplanetary brotherhood, of which George Adamski was the bravest messenger".[13] Desmond Leslie expressed his admiration for Adamski's character by saying: "Of all the contactees, Adamski attracted the most controversy and odium; and none but a man of his strength of character could have survived the onslaught..."[14]

It was Ufologist Richard Heiden to whom Ray Stanford wrote in September 1976 about what Adamski had allegedly told him during his visit in 1958 (or 1956) about "bootlegging" and "flying saucer crap". When Mr Heiden, who can not be mistaken for a fan of Adamski's, contacted the Bureau of Alcohol, Tobacco and Firearms in 1977 about Adamski's boasting of distilling wine under the guise of a religious order, the answer was that no license to manufacture alcohol had been granted to either the Royal Order of Tibet or to George Adamski. Also, Prohibition, the constitutional ban on the production, importation, transportation, and sale of alcoholic beverages in the USA, was in effect from January 1920 until April 1933. Although we now know the Royal Order of Tibet was already active in 1932, it operated from a studio apartment in Los Angeles and a suite or conference room in a Pasadena hotel, neither of which seem terribly well suited for the setup required for a bootlegging operation. The Temple of Scientific Philosophy in Laguna Beach only opened in 1934, after Prohibition had ended.

Nevertheless, the same – long disproved – claim has continued to be used as evidence of Adamski being a conman by subsequent detractors. Indeed, in an online comment from June 2016 Mr Heiden himself insists: "I don't question that Adamski might have spoken as quoted, but it would be just part of his deception and bluster. Maybe he was trying to play down the philosophical/religious side of the Royal Order of Tibet, which might have potentially harmed his 'flying saucer business'."[15] In other words, if one allegation falls flat, we'll come up with another – we'll do anything but take his work seriously because that would show it has been consistent throughout.

The attacks on George Adamski stem from a wide range of motives. Some have the distinct hallmark of a disappointed 'devotee' starting out as an Adamski enthusiast, even translating his work, only to end up with a life-long obsession trying to prove their former idol was a liar. Such an about-turn is often caused by deep disappointment, which history has shown is not uncommon among those who saw their expectations of personal favours unfulfilled.

On the other end of the spectrum are attacks that cross the border into the unethical. Often seen and quoted as the go-to authority on the cultural understanding of the UFO phenomenon and the notion of extraterrestrial contact, Jacques Vallee, in his much-touted 1979 'classic' *Messengers of Deception* says, without reservation: "Only free speculation can open the door to an adequate understanding of what is happening around us."[16] When the thunder of bafflement from reading this statement has subsided, we finally realize how it is possible that so much misunderstanding has taken root as accepted hypotheses in the field of Ufology!

Just a few pages prior to what would elsewhere be the surest way to get one's work dismissed, Mr Vallee offers some revolting conjectures regarding Adamski: "This history of the interaction between flying-saucer contact and politics goes way back, to the early California contactees. In those days many occult groups linked to power-hungry organizations were extremely active. (…) According to some of my own informants, contactee George Adamski had prewar connections with American fascist leader William Dudley Pelley, who was interned during the war. Another seminal contactee, George Hunt Williamson (…), was associated with Pelley's organization, 'Soulcraft,' in the early fifties. In fact, Pelley may have put Williamson in touch with Adamski. Other associates of Williamson during the great era of the flying saucers were such contactees as John McCoy and the two Stanford brothers, Ray and Rex."[17]

Anyone with a fleeting knowledge of political movements knows that fascism is the ultimate 'divide-and-conquer' approach to government, usually based on odious notions of groups or races of people being superior or inferior. Even before *The Invisible Ocean* resurfaced, those with a genuine desire to know what Adamski's world view was would have known it from *Wisdom of the Masters of the Far East*, where he answers the question "What is the law of cosmic brotherhood?": "As God is everything in the universe and manifestations differ only in forms and degrees of manifestations, cosmic brotherhood would have to be an unchangeable, indisputable fact. There is only one cause, one Father."[18]

Actual research by Michel Zirger shows that the only "link" that can be established between Adamski and Pelley is a reply from Adamski that was published in one of the Pelley's periodicals. Zirger concludes: "Contrary to the rumor spread by Jacques Vallee, the very content of the letters published in *Valor* of August 1953, PROVE that neither Adamski nor [his associate Lucy] McGinnis had any contact whatsoever with Pelley BEFORE [Adamski's contact experience in November 1952]."[19]

Thus, Mr Vallee's attempt to besmirch George Adamski's character by association, while completely disregarding his view on life, is nothing less than malicious. And even without Mr Zirger's research it would be an entirely untenable attempt, given that from 1952 until his death in 1965 Adamski was under FBI scrutiny for suspicion of communist sympathies, and if he had indeed held such, the communist block was instrumental in defeating the Nazis in the 1940s. Meanwhile, in the 1950s, when the world was in the middle of an ideological standoff between capitalism and communism, Adamski was telling his audiences that the Space Brothers do not support any specific form of society on Earth: "Such support would be complying with our custom of divisions. They recognize no false divisions of any kind."[20] As it stands, Mr Vallee's "free speculation" – indeed fact-free – only helped keep the door open to a universe of baseless attacks on Adamski's character, adding a whole new layer of meaning to the title of his own book.

Before sceptics now turn to the allegations in the *O.E. Library Critic*'s review of *The Invisible Ocean* as 'evidence' that Adamski was only in it for the money, and point to the acquisition of the property in Laguna Beach as 'proof', it should be noted that the $50,000 purchase was made by Mrs Maud Johnson. Mrs Johnson (née Shlaudeman) was a member of the Decatur, Illinois high society who had moved to South California where she had "become a prophet, enjoying a wide prestige for her writings and lectures under the name of Lalita" and founded the Order of Loving Service in 1933. She was also a disciple of Baba Premananda Bharati, an Indian who had come to American shores to spread Krishna Consciousness. In a review of her book *Square*, the *Decatur Herald* writes: "Lalita is connected with two orders that function in Laguna, 'The Order of Loving Service' and 'The Royal Order of Tibet', this latter, she says, 'being connected with the monasteries of Tibet which have a monk here as representative, and others coming.' "[21]

Note, also, how the $1.5 million estimate for the projected additions to the Laguna Beach property (p.48) had become $2 million through the *O.E. Library Critic*'s grapevine, while official records show that the additions never materialized (p.50). Further, records show that in October 1935 the property was sold to Mrs Marguerite H. Weir, who later that year established Universal Progressive Christianity as a 'parent' organisation for the Royal Order of Tibet (p.52). In other words, George Adamski never owned 758 Manzanita Drive.

Although Adamski's early plans with the monastery and Kashmir-La (p.55) may have seemed ambitious, his actual lifestyle, especially after leaving the Laguna Beach property, shows nothing but modesty. To wit, Desmond Leslie's amazement speaks volumes about Adamski's attitude towards money when he relates the latter's reply to his request to use the photographs of flying saucers, before it was decided to publish their accounts together as *Flying Saucers Have Landed*: "He replied by sending me the whole remarkable set of pictures, along with permission to use them for free. I thought what an extraordinary man. He takes the most priceless pictures of all time and wants no money for them."[22]

Tony Brunt's research confirms this: "Adamski was perfectly placed to make the sacrifices that his unique role involved... Financially he had nothing to lose. Adamski owned no real estate, at least until he left Palomar and moved to a house in Carlsbad as he approached the age of 70. His UFO books brought in some money, but generally he lived on a tight budget."[23]

Hence, while claims that Adamski "was only in it for the money" abound, there is no evidence to be found in his lifestyle, his bank account, or his estate that he ever made a fortune to speak of, and the evidence that is available all points in the opposite direction – that his work as a teacher and harbinger of the Space Brothers was done only as a service for the betterment of all. As he himself often told his audiences: "My advice to you is to help as many as you can – the greater the number you serve, the greater will be the understanding of yourself. This should be the motive of everyone who desires to fulfill the destiny for which he was born."[24]

In *The Invisible Ocean* George Adamski explains: "The fish that lives at the bottom of the ocean cannot understand what is on the top. It is trying to obtain a higher consciousness but being unable to go higher, it does not as yet know the consciousness of the surface. Whatever partakes of that vibration lives in that certain strata." (p.4)

The correspondence with the Ageless Wisdom teaching on consciousness, in the words of H.P. Blavatsky, is striking: "Whatever plane our consciousness may be acting in, both we and the things belonging to that plane are, for the time being, our only realities. But as we rise in the scale of development, we perceive that in the stages through which we have passed we mistook shadows for realities, and that the upward progress of the Ego [i.e. the soul] is a series of progressive awakenings, each advance bringing with it the idea that now, at last, we have reached 'reality'; but only when we shall have reached the absolute Consciousness, and blended our own with it, shall we be free from the delusions produced by Maya [illusion]."[25]

Here we see how our stage in the process of the evolution of consciousness impacts our sense of reality. We can perceive that which falls within our range of perception and understanding. That which lies above it, that resides in higher spheres of vibration or frequencies, will inevitably look unintelligible, and without proper understanding people will mock it and try to bring it down by distorting it.

This is not to disqualify those of us living in strata of slower vibration, but merely disqualifies our ability and suitability to correctly perceive, understand and appreciate what lies above our current level of perception. This principle applies to everyone, because we are all forever engaged on the upward path of growing in conscious awareness. As Adamski puts it in *The Invisible Ocean*: "This invisible ocean of ours is so great that we are like grains of sand in it. It has no limits. No matter how high man goes into consciousness, it will still have no limits for him." (p.10) Again, systems science agrees, as in the words of professor Laszlo: "Consciousness associated with the body is but one phase (...), and consciousness evolves in all phases of its existence."[26]

As we grow in conscious awareness and perception, things that we have outgrown will settle at the bottom of our range of understanding, and we don't feel the need to disrespect those for whom these matters still represent the highest truth, while we ourselves acclimatize to a subtler environment and develop tolerance and respect towards others through a more inclusive view of life.

Until that time arrives for current unbelievers, it will be of little use to try and quell every allegation levelled at those whom history will eventually show to be pioneers. As the above examples make clear, when you disprove one, it will already have been copied in several other places, which will subsequently be quoted as 'proof' by members of the next generation of detractors who have yet to expand their view beyond what they can see or verify with their own eyes, or experience for themselves.

Adamski is certainly not the first or the only teacher of humanity who has met with scorn of this magnitude. As a pioneer herself, Madame Blavatsky suffered very much the same malignant attempts on her character. In this regard, it should be remembered that the report from 1886 which for many decades served to justify the accusations of her alleged fraudulent activities, was unequivocally revoked by the Society for Psychical Research a full hundred years after it had first commissioned and published its report. In a press release dated 8 May 1986 the SPR announced: "Madame Blavatsky, Co-Founder of the Theosophical Society, Was Unjustly Condemned, New Study Concludes."[27] As science is beginning to provide more and more corroboration of Blavatsky's previously esoteric notions about the nature of life, space and evolution (see p.91), we should see her work entering mainstream acceptance sooner rather than later.

Here, again, we see that groundbreaking ideas will first meet with general disbelief, ridicule and rejection because they fall outside the range of perception of the majority of people, before scientific proof and the personal experience of a growing number of individuals begin to clear the way toward general acceptance. If nothing else, this underscores the truth in Adamski's teaching about life in the various strata of the "invisible ocean" as an allegory for the evolution of consciousness.

As we have seen, no-one could maintain that Adamski had no flaws or never made mistakes. But those who limit their efforts to mindlessly repeating unfounded or previously invalidated accusations, ignore historical contexts and deny or lack the understanding of a broader reality than their own should be the last to question his motives.

To be sure, the current writer does not claim to see as much or as far as Madame Blavatsky or George Adamski, but I do know parallels when I see them, both in teachings and in attempts to vilify or discredit them. So let us follow Blavatsky's example who, in the closing words of the Preface to her magnum opus, *The Secret Doctrine*, says: "It is written in the service of humanity, and by humanity and the future generations it must be judged. Its author recognizes no inferior court of appeal. Abuse she is accustomed to; calumny she is daily acquainted with; at slander she smiles in silent contempt."[28]

References

1   Will Johnson (2012), 'George Adamski'. Available at <sites.google.com/site/ theosophyhistory/george-adamski> [Accessed 9 March 2019], and Richard W. Heiden (2016), comment on a post in the UFO Collective Google Group, dated 30 August 2016. Available at: <groups.google.com/forum/#!topic/ufo-collective/TRu5u-mFgzA> [Accessed 9 March 2019]

2   Henry Dohan (2008), *The Pawn of his Creator*, pp.35-36

3   Mother Lalita (n.d.), *The Sacred Symbol*, as reprinted in *Meher Baba Journal*, Vol.4, No.3, January 1942, p.159

4   Lou Zinnstag and Timothy Good (1983), *George Adamski – The Untold Story*, pp.65-66

5   H.P. Blavatsky (1888), *The Secret Doctrine*, 6th Adyar ed. 1971, Vol.V, footnote p.444

6   Jerome Clark (1978), 'Startling New Evidence in the Pascagoula and Adamski Abductions'. In: *UFO Report*, vol.6, no.2, August 1978, p.2. As quoted in Timothy Good (1998), *Alien Base – Earth's Encounters with Extraterrestrials*, p.148

7   Desmond Leslie and George Adamski (1970), *Flying Saucers Have Landed*, Revised and Enlarged Edition, pp.241-42

8   Leslie (1965), 'Obituary: George Adamski', *Flying Saucer Review*, Vol. 11, No.4, July/August, p.18

9   Tony Brunt (2010), *George Adamski – The Toughest Job in the World*, p.13

10 Leslie and Adamski (1970), op cit, p.242

11 Ibidem, pp.242-43

12 Good (1998), op cit, p.155

13 Gerard Aartsen (2010), *George Adamski – A Herald for the Space Brothers*, p.xi

14 Leslie and Adamski (1970), op cit, p.262

15 Heiden (2016), see note 1

16 Jacques Vallee (1979), *Messengers of Deception*, 2008 ed., p.221

17 Ibid., pp.216-17

18 Adamski (1936), *Wisdom of the Masters of the Far East*, pp.58-59

19 Michel Zirger (2017), *"We Are Here!" Visitors Without a Passport*, pp.269-270

20 Adamski (1957-58), *Cosmic Science for the promotion of Cosmic Principles and Truth*, Series No. 1, Part No.1, Question #6

21 *The Decatur Herald*, 15 August 1934, p.6

22 Leslie (1965), op cit

23 Brunt (2010), op cit, p.33

24 Adamski (1962), 'Positive and Negative Thinking'. *Cosmic Science* newsletter Vol.1, No.2, February, p.4

25 Blavatsky (1888), op cit, Vol.I, p.113

26 Ervin Laszlo Ph.D. (2017), *The Intelligence of the Cosmos*, p.42

27 Society for Psychical Research Press Release, 8 May 1986. As referenced in Sylvia Cranston (1993), *HPB – The Extraordinary Life & Influence of Helena P. Blavatsky*, p.xvii

28 Blavatsky (1888), op cit, Vol.I, p.9

CAN ASTROPHYSICS SEE
BEYOND ITS OWN LIMITATIONS ?

This essay was originally written in August 2016, and reworked in 2019. It is included here to underscore how the scientific paradigm is edging ever closer to the recognition that "consciousness is causal, and physical reality its manifestation", and serves as an illustration of how "clusters of vibration" exist at different strata in "the invisible ocean" as vehicles for the manifestation of consciousness.

Despite many today convinced that Earth could not possibly be the only planet to harbour life in the vastness of space, science is nowhere closer to finding evidence for intelligent extraterrestrial life, let alone anyone visiting our planet, as claimed since the 1950s by many who say they were contacted by human-looking beings in highly advanced space craft.

As a result, speculations about the nature of extraterrestrial life range from "extremely tiny super-intelligent explorers"[1] to *Avatar*-inspired 'blue avians'.[2] No surprise, then, that the apparent absence of solid evidence helps to perpetuate the prevailing notion in mainstream society that life on Earth is a space oddity until invaded otherwise. Indeed, without tangible proof for those who have not had access to the alleged 1947 Roswell wreckage even eminent scientific minds seem vulnerable to speculation. For instance, professor Stephen Hawking saw fit to warn NASA not to attempt contact with extraterrestrials[3], prematurely assuming the authority to conclude that all manifestations of life must necessarily be ridden with the same destructive tendency towards competition and greed that is holding humanity back from taking the next step in the evolution of our international community.

Although professor Hawking was undeniably among the sharpest minds in his field, could there be evidence in areas outside physics and cosmology? Could it be that extraterrestrial life is visiting in plain sight, but that the evidence may only be gleaned across a wider spectrum of disciplines? Have we been looking for 'alien' life, perhaps not in the wrong places, but simply not in enough places?

"At the moment, we don't know what more than 90% of the universe is made of," said Mauro Raggi, researcher at the Sapienza University of Rome, in September 2018.[4] And as early as the 1880s Lord Kelvin theorized that "many of our stars, perhaps a great majority of them, may be dark bodies".[5] Astrophysical calculations about the mass of the Universe have since necessitated and spawned one hypothesis after another to explain the fact that modern-day science cannot account for all but about four per cent of its estimated total mass, with dark matter and dark energy the most widely accepted among these.

Perhaps it is good to remind ourselves that scientific rigour through a strictly reductionist approach does not guarantee that we will find tangible evidence, especially regarding a phenomenon – extraterrestrial life – that has proven so elusive to even modern science. As quantum physics pioneer Werner Heisenberg said: "We have to remember that what we observe is not nature in itself, but nature exposed to our method of questioning."[6] It is not that the reductionist approach is false, but rather that it is increasingly proving to be incomplete and therefore inadequate to the task.

So, in order to attempt a better understanding of the whereabouts of intelligent extraterrestrial life, we might expand our research with additional methods of questioning. As scientists themselves admit they don't know where most of the universe may be found, we should remember that several discoveries which point towards more subtle planes of matter have so far been ignored, not only by mainstream science, but also by those who are looking for evidence of extraterrestrial visitors.

To summarize: Based on his experiments British biologist Rupert Sheldrake says that dense physical forms may be seen as the precipitation of the 'blueprints' that exist on subtler levels, according to his theory of "morphogenetic fields" – a sort of memory bank from which Nature retrieves its various solid physical forms.[7] Before Sheldrake, the Austrian doctor Wilhelm Reich experimented with what he called 'orgone radiation', first theorized by German biologist Kammerer as a primordial life force "which is neither heat, electricity, magnetism, kinetic energy (...) nor a combination of any or all of them, but an energy which specifically belongs only to those processes that we call 'life'. That does not mean that this energy is restricted to those natural bodies which we call 'living beings'..."[8] Like many trailblazers, Reich was persecuted, his books were burnt in 1956 and he was sent to prison in 1957 where he died that year. Interestingly, in the 1940s the Soviet inventor and researcher Semyon Kirlian had already developed a technology to photograph the energy fields surrounding living entities, which was later further developed to record human auras, now known as Kirlian photography.[9]

Here we may again quote Werner Heisenberg: "[T]he atoms or elementary particles themselves are not real; they form a world of potentialities or possibilities rather than one of things or facts."[10] Curiously, on the same theme the founder of modern-day esotericism, H.P. Blavatsky, wrote in her seminal work *The Secret Doctrine* (1888): "It is on the doctrine of the illusive nature of matter, and the infinite divisibility of the atom, that the whole science of Occultism is built."[11] (Note: 'Occultism' –from 'occult' = 'hidden'– is used here to mean the science of the energies behind the evolutionary process.)

The notion of planes of subtle matter above the solid, liquid and gaseous that our current science recognizes, has always been a key concept in humanity's ancient Wisdom teachings, which have been gradually reintroduced to the modern world since 1875 in various stages.[12] According to these teachings, these so-called etheric physical planes of matter consist of sub-atomic particles that vibrate at various frequencies, analogous to H_2O molecules vibrating at different frequencies in ice, water and vapour. The experiments mentioned here, in conjunction with science's self-declared nescience, should give us sufficient reason to take the possibility of subtle planes of 'etheric matter' seriously.

As if to show the connection between those experiments and the Wisdom teachings, in 2015 scientific findings were reported which "suggest that dark matter is another kind of sub-atomic particle, possibly forming a parallel universe of 'supersymmetry' filled with supersymmetrical matter that behaves like an invisible mirror-image of ordinary matter."[13] And in December 2018 Dr Jamie Farnes from the Oxford e-Research Centre, Department of Engineering Science proposed a new model of the Universe, that seems to point in the same direction: "...dark energy and dark matter can be unified into a single substance, with both effects [negative masses and matter creation] being simply explainable as positive mass matter surfing on a sea of negative masses."[14]

These findings seem to reflect what US 'contactee' Howard Menger claims he was told by his contacts from space: "Nothing we see with our physical eyes is Truth, but simply a reality in the dimension of a reflection, or an effect, secondary in nature related to a Cause from a primary Source."[15]

It should also be noted here that astrobiologists are finding more and more evidence that life, long thought the result of a chemical accident on an insignificant planet in the backwaters of the Milky Way, is actually rather abundant. For instance, commenting on the recent discovery of traces of micro-organisms dating back to Earth's infancy, Dr Abigail Allwood, of NASA's Jet Propulsion Laboratory, provided a scientific underpinning for the notion that 'life' is not quite as unique as previously thought: "Earth's surface 3.7 billion years ago was a tumultuous place, bombarded by asteroids and still in its formative stages. If life could find a foothold here, and leave such an imprint that vestiges exist even though only a minuscule sliver of metamorphic rock is all that remains from that time, then life is not a fussy, reluctant and unlikely thing. Give life half an opportunity and it'll run with it. (...) Suddenly, Mars may look even more promising than before as a potential abode for past life."[16]

Around the same time the Swiss scientific magazine *Life* reported that, given adequate time and habitat, life will more likely than not evolve into complex forms wherever it occurs,[17] which supports the notion of the predictability of evolutionary outcomes (aka evolutionary convergence).[18] Simply put, this means that the evolution of life, regardless where it occurs, is disposed to result in similar forms. This in turn would lend a scientific underpinning for the claims of people who say they were contacted by human-looking space visitors.

When the first measurements of the surface temperature on Venus came in in 1958, the most famous and most derided of these, George Adamski, was quickly ridiculed for his assertion that his contacts hailed from that and other planets in our solar system. He was not alone, though, in maintaining his claims: Brazilian physicist-contactee Dino Kraspedon (pseudonym for Aladíno Felíx), Canadian researcher-contactee Wilbert Smith, Italian journalist-contactee Bruno Ghibaudi, and US contactees Howard Menger and Buck Nelson all said, more or less publicly, that the spacecraft and their occupants originated from within the solar system, mainly from Mars, Venus, Saturn and a few other planets.

Since the character assassination on the first contactees of the 1950s hardly anyone has placed the origins of the visitors from space within our own solar system. However, UK esotericist Benjamin Creme always maintained: "All the planets of our system are inhabited..." but, he added, "if you were to go to Mars or Venus you would see nobody because they are in physical bodies of etheric matter."[19] This underscores what contactee Howard Menger was told in the 1950s: "[I]f an Earth man in physical body could go there he probably would not see some of the life forms which vibrate more rapidly than his own – no more than he can see the spiritual life forms in and around his own planet. Unless his physical body were processed and conditioned, he could not see the beings on another planet."[20]

Now could the acknowledgement of the abundance of life, and the expanded view of life that presents itself in the corroborations provided here, help to open our minds to the possibility that 'life' does not depend on solid physical forms of expression only and that perhaps it doesn't always – or perhaps rarely – precipitate onto the dense physical plane as it has on our planet? Or else, that life and civilizations on other planets may have evolved beyond the particular rate of vibration that falls within our range of vision? This might also explain the sudden appearance or disappearance of craft as their ability to drop into or out of our range of vision by temporarily lowering the rate of vibration of the sub-atomic particles of their craft or their bodies.

Howard Menger describes how he witnessed a saucer drop into our range of vision: "The ship took the form of a pulsating, fluorescent light, changing in colors from white to green to red. [...] When it was within a foot of the ground and about a hundred feet from the car, it hovered, and I recognized the familiar bell shape. The pulsating colors stopped, it gave off an eery, bluish light, and then portholes appeared."[21]

Reversing the same notion, George Adamski was told, in fact, that the space visitors "can increase the frequency of the activated area of a ship to the point of producing invisibility. Except for our own precaution, your planes could fly blindly into our ship without seeing it. If we permitted you to come as close as that, when you hit, you would find our craft as solid as though functioning in a lower frequency."[22]

If the craft of the visitors from space are in etheric physical matter that would preclude the possibility that anyone is abducted for gene harvesting, hybrid breeding, implanting devices, and other atrocities that are often ascribed to extraterrestrial visitors. In fact, Benjamin Creme says: "Nobody is ever taken up in a spaceship in a physical body. It is impossible. These spaceships are not solid physical. To be taken up into a spaceship you have to be taken out of the dense physical body and you go in the etheric [body] into the spaceship, which are in themselves etheric. It is still physical, but etheric physical."[23]

Interestingly, a possible description of the process of being taken out of the (dense physical) body also comes from Howard Menger, who describes how a bluish beam was aimed at him when he was allowed to board a ship: "...as it struck my head I felt a tingling sensation, warm, and rather pleasant. I stood in my tracks as he slowly played the beam downward over my body until it had reached my feet."[24] His hosts later explained: "We projected the beam on you to condition and process your body quickly so you could enter the craft. What actually happened was that the beam changed your body frequency to equal that of the craft."[25]

The accounts of various contactees also present tacit evidence of their having been taken out of the physical body as they testified to a heightened state of awareness once they were on a ship.

American contactee Orfeo Angelucci seems to give some impression of what it must feel like once one is in this 'unearthly' state: "The interior was made of an ethereal mother-of-pearl stuff, irridescent [sic] with exquisite colors that gave off light... There was a reclining chair directly across from the entrance. It was made of the same translucent, shimmering substance – a stuff so evanescent that it didn't appear to be material reality as we know it... As I sat down I marveled at the texture of the material. Seated therein, I felt suspended in air, for the substance of that chair molded itself to fit every surface or movement of my body. As I leaned back and relaxed, that feeling of peace and well-being intensified."[26]

The similarities of Angelucci's descriptions with those of other contactees are striking. For instance, Italian contactee Giorgio Dibitonto writes about his experience in 1980: "The central room was illuminated with light that seemed to come from all directions, as no single light source was to be seen. ... An unaccustomed empathy prevailed; we were all flooded with this same unearthly light, and with an energy that was more spiritual than physical."[27] Likewise, Howard Menger noted: "[T]he walls became brighter, as if illumined somehow from inside themselves..."[28]

George Adamski described the interior of a disk as follows: "Within the craft there was not a single dark corner. I could not make out where the light was coming from. It seemed to permeate every cavity and corner with a soft pleasing glow. There is no way of describing that light exactly. It was not white, nor was it blue, nor was it exactly any other color that I could name."[29] And scientist Michael Wolf's description of the interior of a saucer is also strongly reminiscent of Adamski, Angelucci and Dibitonto's accounts: "We were standing in what seemed to me as a very familiar room, brilliantly lit, but not as much as to hurt the eyes. The light seemed not to emanate from any single source, but was everywhere."[30]

Since the relentless disinformation campaign that was initiated in the mid-1950s to defuse and ridicule the information about the need for international cooperation coming from the original contactees in the midst of the Cold War[31], many dismiss the information that contactees claim they were given. The unwavering, but undeserved faith in a strictly reductionist approach to scientific research has caused many to treat the postulates of the Wisdom teachings with equal disdain.

But when we broaden our inquiry into the nature of extraterrestrial life by including additional methods of questioning, as we take into account science's self-declared limitations, avantgarde scientific experiments that look beyond those limitations, the latest scientific insights, previously esoteric notions about the nature of matter, and eyewitness accounts from people the world over, a plausible window opens onto a world as yet unseen, but one that can well be explained and argued, even within the limits of our current understanding.

So, when next we read scientific assertions that there is no life on the other planets in our solar system, perhaps we should add "...on the dense physical planes of matter".

Note: Readers who think the ET hypothesis as an explanation for UFOs remains far-fetched or unsupported by credible sources, or who have difficulty taking accounts of the 1950s contactees seriously, may find it of interest to see how such claims have since been corroborated by officials and scientists, in a brief video compilation by the author: 'End the UFO/ET disclosure nonsense!' (available on YouTube).

References

1 Silvano P. Colombano at NASA Ames Research Center, as quoted in Chelsea Ritschel (2018), 'Aliens may have already visited Earth, NASA Scientist Says'. *The Independent*, 4 December. Available at <www.independent.co.uk/life-style/gadgets-and-tech/aliens-earth-nasa-scientist-space-extraterrestrial-travel-seti-a8667506.html>. [Accessed 5 December 2018]

2 The earliest mention of 'Blue Avians' appears in a *Stillness in the Storm* blog (2 March 2015), and seems to be a conflation of ancient Native American myth and modern Hollywood fantasy. Available at <sitsshow.blogspot.nl/2015/03/benjamin-fulford-and-david-wilcocks.html>. [Accessed 4 September 2016]

3 'Stephen Hawking warns over making contact with aliens'. *BBC News* online, 25 April 2010. Available at <news.bbc.co.uk/2/hi/8642558.stm>. [Accessed 5 December 2018]

4 Ian Sample (2018), 'Scientists hunt mysterious 'dark force' to explain hidden realm of the cosmos'. *The Guardian*, 3 September. Available at <www.theguardian.com/science/2018/sep/03/scientists-hunt-for-dark-force-to-discover-what-the-universe-is-made-of>. [Accessed 4 September 2018]

5 Stephanie M. Bucklin (2017), 'A History of Dark Matter'. *Ars Technica*, 3 February. Available at <arstechnica.com/science/2017/02/a-history-of-dark-matter/>. [Accessed 5 December 2018]

6 Werner Heisenberg (1958), *Physics and Philosophy: The Revolution in Modern Science*, p.25

7 Rupert Sheldrake (1981), *A New Science of Life – The Hypothesis of Formative Causation*

8 Wilhelm Reich M.D. (1960), *Selected Writings*, p.195

9 J.D. Zakis, B.L. Lithgow, I. Cosic, J. Fang (1999), 'Kirlian Photography and Techniques'. Monash University, Clayton, Victoria, Australia. Available at <www.eng.monash.edu.au/non-cms/ecse/ieee/ieeebio1999/p160.htm10>. [Accessed 10 December 2018]

10 Heisenberg (1958), op cit, p.186

11 H.P. Blavatsky (1888), *The Secret Doctrine*, 6th Adyar ed. 1971, Vol.II, p.244

12 For an annotated overview, see Gerard Aartsen (2008), 'Our Elder Brothers Return – A History in Books (1875 – Present)'. Available at <www.biblioteca-ga.info>

13 Steve Connor (2015), 'The galaxy collisions that shed light on unseen parallel Universe'. *The Independent*, 26 March. Available at <www.independent.co.uk/news/science/the-galaxy-collisions-that-shed-light-on-unseen-parallel-universe-10137164.html>. [Accessed 27 March 2015]

14 'Bringing balance to the universe: New theory could explain missing 95 percent

of the cosmos'. Phys.org, 5 December 2018. Available at <phys.org/news/2018-12-universe-theory-percent-cosmos.html#jCp>. [Accessed 10 December 2018]

15 Howard Menger (1959), *From Outer Space to You*, p.173

16 Ian Johnston (2016), 'World's oldest fossils found in discovery with "staggering" implications for the search for extraterrestrial life'. *The Independent*, 31 August. Available at <www.independent.co.uk/news/science/oldest-fossils-world-alien-life-earth-mars-greenland-a7218191.html>. [Accessed 2 September 2016]

17 William Bains and Dirk Schulze-Makuch (2016), 'The (Near) Inevitability of the Evolution of Complex, Macroscopic Life'. *Life* magazine, Vol.6, Issue 3, 30 June. Available at <www.mdpi.com/2075-1729/6/3/25/htm>. [Accessed 12 August 2016]

18 Matthew Wills (2016), 'What do aliens look like? The clue is in evolution'. *The Conversation*, 19 August. Available at <theconversation.com/what-do-aliens-look-like-the-clue-is-in-evolution-63899>. [Accessed 2 September 2016]

19 Benjamin Creme (2001), *The Great Approach – New Light and Life for Humanity*, p.129

20 Menger (1959), op cit, p.162

21 Ibid., p.74

22 George Adamski (1955), *Inside the Space Ships*, p.156

23 Creme (2010), *The Gathering of the Forces of Light – UFOs and their Spiritual Mission*, p.49

24 Menger (1959), op cit, p.83

25 Ibid, p.84

26 Orfeo Angelucci (1955), *The Secret of the Saucers*, pp.20-21

27 Giorgio Dibitonto (1990), *Angels in Starships*, p.67

28 Menger (1959), op cit, p.83

29 Adamski (1955), op cit, p.50

30 Michael Wolf (1996), *The Catchers of Heaven*, p.199

31 As documented in Aartsen (2015), *Priorities for a Planet in Transition*

GEORGE ADAMSKI
BIBLIOGRAPHY

Reproduced from Gerard Aartsen (2010), *George Adamski – A Herald for the Space Brothers*, with latest corrections and additions.

Books

Answers to Questions Most Frequently Asked About Our Space Visitors And Other Planets
Palomar Gardens, CA, USA : G. Adamski, 1965. (30 p.)
[Selected questions and answers (revised) from *Cosmic Science* bulletin Series No.1, Parts 1-5 (1957-1958).]

Cosmic Philosophy
San Diego, CA, USA : G. Adamski, 1961. (iv, 87 p.)
Republished as:
Cosmic Philosophy
Freeman, SD, USA : Pine Hill Press, 1972.
Pt Pleasant, WV, USA : New Saucerian, 2018 [list date 1961] (96p.)

Flying Saucers Farewell
New York, NY, USA; London, UK : Abelard-Schuman, 1961. (190 p.)
Re-published as:
[The Strange People, Powers, Events] Behind the Flying Saucer Mystery
New York, USA : Warner Paperback Library, 1967, 1974. (159 p.)
[The Extraordinary People, Powers, Events] Behind the Flying Saucer Mystery II
CA, USA : George Adamski Foundation, 2016. (244 p.)
Behind the Flying Saucer Mystery: Ancient Astronauts, The Space Brothers, and The Silence Group
Clarksburg, WV : Saucerian Press, 2018 [list date 1967]. (160 p.)

Flying Saucers Have Landed [with Desmond Leslie]
New York, USA : The British Book Centre, 1953, 1954, 1955, 1956, 1967, 1972. (232 p.)
London, UK : T. Werner Laurie, 1953, 1954, 1959. (232 p.)

London, UK : Panther Books, 1957.
New York, NY, USA : Warner Paperback Library, 1967.
London, UK : Tiptree Book Service, 1970. (264 p.)
London, UK : Book Club Associates, 1973.
Re-published as:
Flying Saucers Have Landed. Revised and Enlarged Edition
London, UK : Neville Spearman 1970, 1972, 1976. (281 p.)
London, UK : Futura 1977, 1978.
Flying Saucers Have Landed, The
[No place], USA : New Illuminet Press, 2017. (168p.)

Inside the Space Ships
New York, NY, USA : Abelard-Schuman, 1955, 1957. (256 p.)
Toronto, Canada : Nelson, Foster and Scott Ltd, 1955. (256 p.)
London, UK : Arco Publishers / Neville Spearman, 1956, 1957, 1958, 1966, 1971. (236 p.)
New York City, NY, USA : Fieldcrest, 1966.
Re-published as:
Inside the Flying Saucers
New York, NY, USA : Warner Paperback Library, 1967, 1969, 1973, 1974. (192 p.)
Tokyo, Japan : Asahi Press, 1980 in 2 vols [as English-language textbooks with Japanese annotations].
Inside the Spaceships : Ufo Experiences of George Adamski 1952-1955
Vista, CA, USA : George Adamski Foundation International, 1980. Reprinted 1995, 2014. (296 p.) [Includes Adamski's part from *Flying Saucers Have Landed*.]
Inside the Flying Saucers
[No place], USA : anon., 2014. (106p.)
Inside the Space Ships
[No place], New Zealand: Muriwai Books, 2018. (173p.)

Invisible Ocean, The
Los Angeles, CA, USA : Leonard-Freefield Co., 1932. (23 pp)

Petals of Life : Poems
Laguna Beach, CA, USA : Royal Order of Tibet, 1937. (16 p.)

Pioneers of Space : A Trip to the Moon, Mars and Venus
   Los Angeles, CA, USA : Leonard-Freefield Co., 1949. (260 p.)
Republished as:
Lost Book of George Adamski – Pioneers of Space, The
   New Brunswick, NJ, USA : Global Communications, 2008 (308
   p.) [Includes a reprint of *Wisdom of the Masters of the Far East.*]
Pioneers of Space. South Boston, VA, USA: Black Cat Press, 2010 (207p.)
Pioneers of Space (Kindle ed.). Annotated by Ron Miller (ed.) Wake
   Forrest, NC, USA: Baen Books, 2013. (216p.)
*Pioneers of Space. George Adamski's Lost Tale of a Journey from the
   Moon to Mars and Venus*
   [No place], USA : anon., 2017. (162p.)

Possibility of Life on Other Planets, The
   [No place], anon. 1946. (32 p.)

*Wisdom of the Masters of the Far East : Questions and Answers by the
   Royal Order of Tibet : Vol. 1*
   Laguna Beach, CA, USA : Royal Order of Tibet, 1936. (67 p.)
Re-published as:
Wisdom of the Masters of the Far East
   Mokelumne Hill, California, USA : Health Research, 1974. (67 p.)
   Hastings, Sussex, UK : Society of Metaphysicians, 1986. (67 p.)
   Vista, CA, USA : Science of Life, 1990.
   Pomeroy, WA, USA : Health Research, 2000. (66 p.)
   Pt Pleasant, WV, USA : New Saucerian, 2018. (72 p.)
Also reprinted in *The Lost Book of George Adamski – Pioneers of Space*,
   New Brunswick, NJ, USA : Global Communications, 2008.

Courses

Science of Life Study Course
   [Valley Center, CA, USA : G. Adamski], 1964. (12 lessons)

Telepathy : The Cosmic or Universal Language
   San Diego, CA, USA : G. Adamski, 1958. (Parts I-III; 31, 32,
   42 p.)
   Vista, California, USA : Ufo Education Center/GAF, [197-?].

Bulletins, reports and pamphlets

Challenge to Spiritual Leaders, A
Valley Center, CA, USA : G. Adamski, 1965. (3 p.)
[Reprinted from C.A. Honey (ed.), *Cosmic Science* newsletter,
Vol.1, No.1. January 1962.]

Cosmic Consciousness
[No details. Mentioned in Winfield Brownell, *UFOs, Keys
to Earth's Destiny*, Lytle Creek, CA, USA : Legion of Light
Publications, 1980, p.24.]

*Cosmic Science for the Promotion of Cosmic Principles and Truth :
Questions and Answers*
Valley Center, CA, USA : Cosmic Science, 1957-58. (Parts 1-5;
16, 28, 32, 32, 36 p.)

Gravity and the Natural Forces of the Universe
[Transcript of an informal talk given in Vista, CA, USA, early
1960s.] (3 p.)

In My Father's House Are Many Mansions
Detroit, Michigan, USA : Interplanetary Relations, 1955. (14 p.)
[Transcript of a press conference which Adamski held for a
group of priests in September 1955, plus subsequent questions
and answers.]
Re-published as: *Many Mansions*
Willowdale, ON, Canada : SS&S Publications, 1974, 1983. (20 p.)

Latest Fascinating Experiences
[No details. Listed in R. Michael Rasmussen, *UFO Bibliography : an
annotated listing of books about flying saucers*. La Mesa, CA, USA,
1975.]

Law of Levitation, The
[No details. Mentioned in Winfield Brownell, *UFOs, Keys
to Earth's Destiny*, Lytle Creek, CA, USA : Legion of Light
Publications, 1980, p.24]

Press Conference with Detroit Ministers
Detroit, MI, USA : The Interplanetary Foundation, 1955. (10 p.)

Private Group Lecture for Advanced Thinkers
Detroit, MI, USA : Civilian Flying Saucer Researchers, 1955.
(17 p.) [Transcript of the CFSR's meeting with Adamski, 4 May 1955.]

Religion and Saucers
Detroit, MI, USA : Interplanetary Relations, 1955. (22 p.)
[Transcript of a lecture held in Detroit, 19 September 1955.]

Satan, Man of the Hour
[No place; no page count], The Royal Order of Tibet, 1937.
Reprinted (revised) in *Flying Saucers Farewell*, 1961.

Special Report : My Trip to the Twelve Counsellors' Meeting That Took Place on Saturn, March 27-30, 1962
Vista, CA, USA : Science of Life, 1962. (2 parts, 5p. + 4 p.)
Jane Lew, WV : New Age Books, 1983. (11 p.)

Universal Jewels of Life
[Laguna Beach, CA, USA : Universal Progressive Christianity]
January 1936 – ? (Monthly publication handed out free at Royal Order of Tibet/Universal Progressive Christianity meetings.)

World of Tomorrow
Detroit, MI, USA : Interplanetary Relations, 1956. (19 p.)
[Transcript of a lecture held in Detroit, 20 September 1955.]

Articles/Letters

1937. 'The Kingdom of Heaven on Earth'. [No details. Available on GAF website.]
1950. 'Flying Saucers As Astronomers See Them' [with Maurice Weekly]. *FATE magazine* 3(6) : pp.56-59.
1951. 'I Photographed Space Ships'. *FATE magazine* 4(5): pp.64-74.
1955. 'Adamski's Answer to Baker'. *Mystic Magazine*, June: pp.96-97

'Time Will Tell', *Saucerian Bulletin* No.6: p.33.

1956. 'Inside a Flying Saucer'. *Real Adventure*, July 1956: pp.40-43, pp.84-97.

1957. 'Letter about GAP'. Reprinted in Daniel Ross (ed.), *UFOs and Space Science*, No.1, December 1989, p.20.

'Flying Saucers versus the Supernatural'. *Flying Saucer Review* 3(5), p.17-19.

1958. 'Adamski Answers Washington Denial Regarding R. E. Straith'. *Flying Saucer Review* 4(4), pp.8-9.

1959. 'Who is Trying to Stop the Truth Coming Out?'. *Flying Saucer Review* 5(1), pp.18-19.

1960. 'Think This One Over'. *Saucers, Space & Science*, No.18, December 1960, pp.11-12

1962. 'World Disturbances'. *Cosmic Science* newsletter Vol.1, No.1, January 1962, p.4.

'Positive & Negative Thinking Vs Motive'. *Cosmic Science* newsletter Vol.1, No.2, February 1962, p.4.

'George Adamski Editorial', *Cosmic Science* newsletter Vol.1, No.3, March 1962.

'Editorial by George Adamski', *Cosmic Science* newsletter Vol.1, No.4, April 1962.

'Many Are Called But Few Are Chosen', *Cosmic Science* newsletter Vol.1, No.9, September 1962, pp.1-2.

'A Challenge to Spiritual Leaders', *Cosmic Science* newsletter Vol.1, No.11, November 1962, pp.1-2.

'George Adamski's Spiritual Crusade For Survival – Results of Nuclear Testing', *Cosmic Science* newsletter Vol.1, No.12, December 1962, pp.2-3.

1963 'Article by George Adamski'. *Cosmic Science* newsletter, September 1963, pp.12-13. [Announcement for the *Science of Life* Study Course.]

1964. 'Authority versus Common Sense'. *Probe* 1964. (Reprinted in Daniel Ross (ed.), *UFOs and Space Science*, No.1, December 1989., p.18-19. This issue also carried the transcript of an undated informal discussion by Adamski, which the editor titled 'Nature is the True Teacher', on pp.16-17.)

'The Space People'. (First published in Gerard Aartsen

(2010), *George Adamski – A Herald for the Space Brothers* (pp.111-126)

[n.d.] 'The Impersonal Deity'. [Reproduced in *UFO Contact – IGAP Journal* Vol.3, No.3, June 1968, pp.83-84]
'Individual Analysis and Thought Control'.
'Transformation of Body Consciousness'.
[Both first published in the current volume, pp.27-38]

Compilations

Adamski Documents : Part One, The [Edited by Gray Barker]
Clarksburg, WV, USA : Saucerian Publications, 1980. (108 p.)
[Reprints and extracts from the author's correspondence, plus
miscellaneous articles. A further compilation of correspondence
to and from George Adamski can be found as Appendix IV in
Lou Zinsstag , *George Adamski – Their Man on Earth* (1990).]
Re-issued as: *Adamski Papers, The*
Jane Lew, WV, USA : New Age Books, 1983. (100 p.)

Book of Adamski [Edited by Gray Barker]
Clarksburg, WV, USA : Saucerian Publications, n.d. [c1966].
(78 p.)
Includes the following texts by George Adamski:
- 'My Fight With the Silence Group' (pp.33-37)
- 'Questions and Answers – Most frequent questions answered
by George Adamski' (selection from *Cosmic Science* bulletin
Series No.1, Parts 1-5, 1957-58) (pp.39-47)
- 'Space Age Philosophy' – Compilation of four articles:
'Positive and Negative Thinking'* ; 'Many Are Called But Few
Are Chosen'*; 'How to Know A Spaceman If You See One'
[1962]; and 'Annihilation' (pp.49-55)]
*Reprinted from C.A. Honey (ed.), *Cosmic Science* newsletter
Vol.1, Nos.2 and 9.
Later reproductions:
Desert Hot Springs, CA, USA : Lizardhaven Press [c1995]
IN, USA : Saucer's Apprentice [c2008]
Pt Pleasant, WV, USA : New Saucerian, 2014 (80 p.)

Cosmic Bulletin
Valley Center, CA, USA : The Adamski Foundation, 1965.
[Series of documents sent by Adamski to his followers from
December 1963, until his passing in 1965.]

Notes from Adamski Lectures (1963-1964)
Valley Center, CA, USA : Science of Life [196-?]. (15 p.)

*We Are Not Alone in the Universe: Six Essays on Flying Saucers and
the Significance of Their Appearance* [with others; edited by S.K.
Maitra] in: *The Sunday Standard*
New Delhi, India : S.K. Maitra, 1958. (25 p.)

FURTHER READING

Other works by Gerard Aartsen:

Online reference:

Our Elder Brothers Return – A History in Books
(1875 - Present)
published at www.biblioteca-ga.info

Books:

Before Disclosure – Dispelling the Fog of Speculation

Priorities for a Planet in Transition –
The Space Brothers' Case for Justice and Freedom

Here to Help: UFOs and the Space Brothers

George Adamski – A Herald for the Space Brothers

The author's books have also been
translated and published in
Dutch, Japanese, French, Spanish,
German and Finnish.
Available from all major online vendors.

FURTHER READING

Alice A. Bailey, *Telepathy and The Etheric Vehicle*
—, *The Consciousness of the Atom*

Tony Brunt, *George Adamski – The Toughest Job in the World*

Sylvia Cranston, *HPB – The Extraordinary Life & Influence of Helena P. Blavatsky*

Benjamin Creme, *The Ageless Wisdom Teaching – An introduction to humanity's spiritual legacy*
—, *The Gathering of the Forces of Light – UFOs and their Spiritual Mission*

Timothy Good, *Alien Base – Earth's Encounters with Extraterrestrials*

Ervin Laszlo, *The Intelligence of the Cosmos*
—, *What is Reality?*

Murdo MacDonald-Bayne, *Beyond the Himalayas*
—, *The Yoga of the Christ*

Mohammed Mesbahi, *The Commons of Humanity*

Vera Stanley Alder, *From the Mundane to the Magnificent*

Michel Zirger, *Authenticating the George Adamski Case*